LabVIEW for Electrical Engineers and Technologists

LabVIEW programming tutorial with practical electrical examples

Stephen P. Tubbs, P.E.
formerly of the
Pennsylvania State University,
currently an
industrial consultant

1344 Firwood Dr.
Pittsburgh, PA 15243

NOTICE TO THE READER

The author does not warrant or guarantee any of the products, equipment, or software described herein or accept liability for any damages resulting from their use.

This publication is an independent publication. National Instruments is not affiliated with the publisher or the author, and does not authorize, sponsor, endorse, or approve this publication.

The reader is warned that electricity and the construction of electrical equipment are dangerous. It is the responsibility of the reader to use common sense and safe electrical and mechanical practices.

Printed in the United States of America and United Kingdom.

ISBN 978-0-9819753-3-7

CONTENTS

1.0 LabVIEW

In LabVIEW Lab is short for **Lab**oratory and VIEW is short for **V**irtual **I**nstrumentation **E**ngineering **W**orkbench.

LabVIEW is a graphical (visual) programming language that acquires data from input devices, files, or manual keyboard entry. It then processes the data and produces output data. Its programming language is named G and its programs are called VIs (**V**irtual **I**nstruments). G can graphically or numerically display the output data or can send the output data through output devices to control equipment. LabVIEW G programs are different from most other language's programs. LabVIEW's G uses graphical input blocks, function blocks, output blocks, and nodes connected by "dataflow wires" rather than lines of text statements like those found in C, FORTRAN, MATLAB, etc. Also, LabVIEW executes each block or node as soon as data is available in all of its inputs rather than executing statements sequentially.

Although LabVIEW can do significant computations, it is best suited for real-time data acquisition, simple data processing, and production of numerical and graphical outputs. One might think of it as a very sophisticated strip-chart recorder. LabVIEW also has real-time control abilities similar to those of programmable controllers.

For small programs, like those shown in this book, G programming can be easier than more conventional programming languages like C, FORTRAN, or MATLAB. However for large programs, graphical programming with G is more difficult. Graphical programs tend to look like spaghetti when they are too large. To remedy this, LabVIEW has "Formula Node" function blocks and "MathScript". A "Formula Node" function block evaluates equations inside its graphical programming block. "MathScript" is a separately-purchased add-on program that allows mathematical programs to be written in text form, as they are in C, FORTRAN, and MATLAB.

The National Instruments Corporation created LabVIEW in 1986 for use on Apple Macintosh computers. LabVIEW is still the property of National Instruments. It is not managed and standardized by a third party like ANSI, C and FORTRAN have been. Now versions of G can be run on Microsoft Windows, Linux, and Macintosh.

VIs (G programs) contain two main components: a "Front Panel" and a "Block Diagram". The "Front Panel" has input and output blocks for inputting values and displaying output values. The "Block Diagram" is the working part of the VI. It shows input, output, and logical function blocks connected together by "dataflow wires". A third type of component that VIs may contain is the "Connector Pane". "Connector Pane"s are used to define sub-VIs, the VI equivalent of subroutines. With a "Connector Pane" a sub-VI is defined that can be called up for use in another VI. Smaller VIs, such as those in this book, do not need "Connector Panes".

National Instruments also offers many specialized LabVIEW add-on modules and toolkits (add-on programs).

Information can be found through the National Instruments Corp. website, http://www.ni.com/labview. The National Instruments Corp. is located at 11500 N Mopac Expwy, Austin, TX 78759-3504. They can be contacted by phone at (888) 280-7645.

This book was written using LabVIEW 2009 free evaluation software and the LabVIEW 2009 Student Edition.

1.1 PURCHASED LabVIEW VERSIONS

LabVIEW will operate on Microsoft Windows, Macintosh, and Linux operating systems.

The following chart details features available with the different versions:

Feature	Base	Full	Professional	Developer Suite
User Interface Development	X	X	X	X
Data Acquisition Functions and Wizards	X	X	X	X
Instrument Control Functions and Wizard	X	X	X	X
Report Generation & Data Storage	X	X	X	X
Calling External Code	X	X	X	X
Modular and Object-Oriented Development	X	X	X	X
Network Communication	X	X	X	X
LabVIEW SignalExpress Included	-	X	X	X
Math, Analysis, & Signal Processing	-	X	X	X
Event Driven Programming	-	X	X	X
Application Distribution	-	-	X	X
Software Engineering Tools	-	-	X	X
Productivity Toolkits Included	-	-	-	X

The following chart gives list prices at the time of printing. Contact National Instruments for current prices.

Operating System	Base	Full	Professional
Microsoft Windows	$1,479	$3,077	$5,090
Macintosh	-	$2,599	$4,299
Linux	-	$2,599	$4,299

Technical support and upgrades are free for the first year after purchase. After that they cost 20% of the current price of the software per year. If technical support and upgrades have lapsed they can be begun again with a payment of 50% of the software's current price.

1.2 LabVIEW 30-DAY FREE EVALUATION SOFTWARE

Free LabVIEW 2010 Evaluation Software is available for the Windows operating system (32-bit and 64-bit) and Mac OS X for 30 days. See http://www.ni.com/trylabview/ for evaluation downloads or DVDs. The evaluation softwares have the functionality of LabVIEW Professional.

With a separate request, most of the LabVIEW modules, toolkits, and device drivers are also available for evaluation for 30 days.

In some cases National Instruments will extend the evaluation period.

VIs (G programs) created with the Evaluation Software that uses the features found only in the Professional version will not operate in the Base and Full versions.

To remove LabVIEW after the evaluation period, LabVIEW, its help files, and its examples must be uninstalled.

The Evaluation Version is the same as the full Professional Version of LabVIEW, except that it expires 30 days after it is launched.

Instructions for the downloading, installation, and start-up of the LabVIEW program can be found on the National Instruments website.

1.3 LabVIEW STUDENT EDITION

According to a National Instruments website: "With the LabVIEW 2009 Student Edition, students can design graphical programming solutions to their classroom problems and laboratory experiments with software that delivers the graphical programming capabilities of the LabVIEW professional version. The Student Edition is also compatible with all National Instruments data acquisition and instrument control hardware. Note: The LabVIEW 2009 Student Edition is available to students, faculty, and staff for personal educational use only. It is not intended for research, institutional, or commercial use."

National Instruments promotes the LabVIEW 2009 Student Edition Textbook Bundle which includes the LabVIEW Student Edition software and Dr. Robert H. Bishop's popular introductory textbook *Learning with LabVIEW*. See reference 1 on page 134.

LabVIEW can be purchased with an academic site license. This provides unlimited installations within a department, college, or campus.

1.4 SYSTEM REQUIREMENTS

SYSTEM REQUIREMENTS must be met for LabVIEW to have all features available. Some LabVIEW features may operate with hardware or software that does not meet all the requirements.

WINDOWS OPERATING SYSTEM

Hardware

- Requires a minimum of Pentium III or Celeron 866 MHz or equivalent
- Recommends at least a Pentium 4/M or equivalent
- Requires a minimum of 256 MB RAM
- Recommends 1 GB RAM or higher
- Recommends at least 1.6 GB of disk space
- Requires a screen resolution of at least 1,024 x 768 pixels
- Requires a minimum color palette setting of 16-bit color

Software

- Requires Windows 7, XP, Vista, or Server 2003 R2 (32-bit)/Server 2008 R2 (64-bit)
- Requires Internet Explorer 5.5 Service Pack 2 or later to view and control a front panel remotely using Internet Explorer
- Requires .NET Framework 2.0 or later to use .NET functions and applications
- Requires Adobe Reader 6.0.1 or later to view or search PDF versions of LabVIEW manuals

MACINTOSH OPERATING SYSTEM

Hardware

All Intel-based Macs (PowerPC processors are not supported)
Requires a minimum of 256 MB RAM
Recommends 1 GB RAM or higher
Recommends at least 1.2 GB of disk space
Requires a screen resolution of at least 1,024 x 768 pixels

Software

Requires Mac OS X 10.5 or later
Requires XCode 2.4.1 or later for LabVIEW Application Builder
Requires Adobe Reader 6.0.1 or later to view or search PDF versions of LabVIEW manuals
Recommends Firefox 1.0.2 or later or Safari 1.3.2 or later to view LabVIEW Help

LINUX OPERATING SYSTEM

Hardware

Intel x86 processors with kernel version 2.2.x, 2.4.x, or 2.6.x
Requires a minimum of Pentium III or Celeron 866 MHz or equivalent
Recommends at least a Pentium 4/M or equivalent
Requires a X Window System server, such as XFree86 or X11R6.org
Requires a minimum of 256 MB RAM
Recommends 1 GB RAM or higher
Requires at least 630 MB of disk space for a minimum LabVIEW installation
Requires at least 947 MB of disk space for a complete LabVIEW installation
Requires a screen resolution of at least 1,024 x 768 pixels

Software

Requires Red Hat Enterprise Linux WS4 or later and open SUSE 11.0 or later
Can run without hardware support on any other distributions that provide GNU C Library (glibc, also known as libc.so.6) version 2.2.4 or later
Recommends that Firefox 1.0.2 or later or Mozilla 1.2 or later are used to view LabVIEW Help
Requires GNU C Library version 2.2.4 or later
Requires Adobe Reader 6.0.1 or later to view or search PDF versions of LabVIEW manuals

1.5 INSTALLING THE 30-DAY FREE EVALUATION LabVIEW

This book covers the basics of "LabVIEW", but does not cover details on LabVIEW's toolkits or modules. At the beginning of the installation process select only "LabVIEW English (Base/Full/Professional)".

During downloading, the installation program asks for a Serial Number. For the 30-day free evaluation LabVIEW this can be left blank.

If the program is being loaded in from a "NI LabVIEW 2009 Evaluation DVD"s, insert DVD 1 and follow the on-screen instructions. The setup program will probably recommend a software patch (about 148 MB) that needs to be downloaded from a NI internet site. Expect the whole installation process to take about an hour.

If the program is being downloaded from the NI website, http://www.ni.com/trylabview/, follow the on-screen instructions. The download is 763 MB.

After files are loaded and unzipped, left-click on the "Measurement & Automation" icon. Then left-click on "My System – Measurement & Instrumentation Explorer" and "My System, and Software". Next right-click on "LabVIEW 2009" and left-click on "Launch LabVIEW 2009". The window shown in Figure 1-5-1 appears.

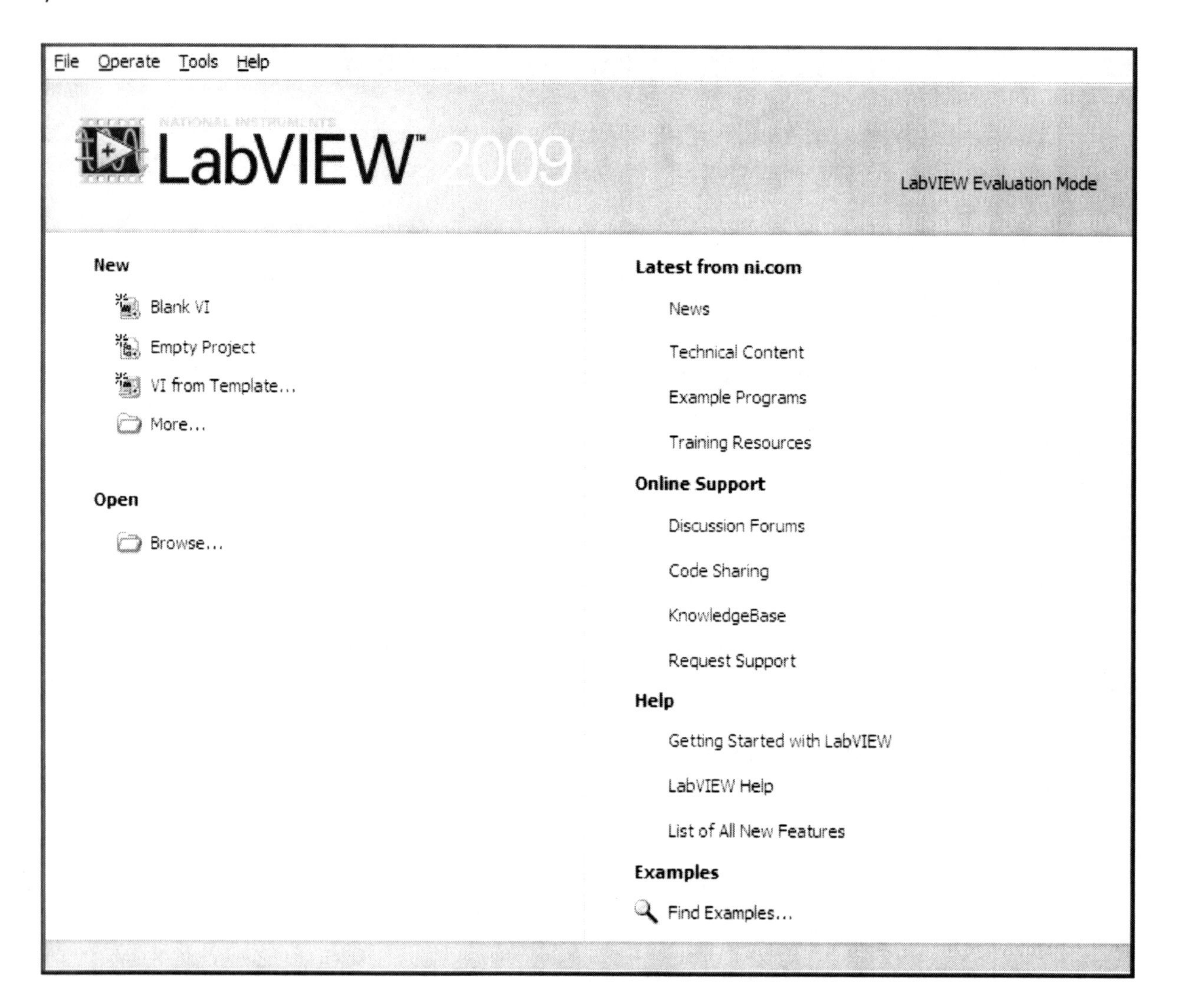

Figure 1-5-1 "Getting Started" window for the "LabVIEW 2009" "LabVIEW Evaluation Mode".

1.6 INSTALLING A DAQmx DRIVER

DAQmx driver software is used by LabVIEW to communicate with DAQ input and output devices. It is required in the examples of Sections 4.3 and 4.4.

It is free. For this book the most recent DAQmx driver was obtained from http://joule.ni.com/nidu/cds/view/p/id/1444/lang/en Then the NI Downloader, "NI-DAQmx 9.0.2 Downloader" was run. This unloaded a 1322 MB file and, when requested, installed it.

Like LabVIEW, the DAQmx driver software is being continuously updated. It is possible that the above version is outdated. Search for the latest on the NI.com website or speak with your NI representative.

1.7 LabVIEW SIGNALEXPRESS

According to a NI website, "National Instruments LabVIEW SignalExpress is interactive, measurement software for quickly acquiring, analyzing, and presenting data from hundreds of data acquisition devices and instruments, with no programming required."

Although related to LabVIEW, LabVIEW SignalExpress is a separate program that is purchased separately from LabVIEW. LabVIEW and LabVIEW SignalExpress can function together or as stand-alone programs.

Information on LabVIEW SignalExpress can be found on http://www.ni.com/labview/signalexpress/.

LabVIEW SignalExpress is not covered in this book.

1.8 LabVIEW-RELATED 30-DAY FREE EVALUATION SOFTWARE

LabVIEW Toolkits
- Adaptive Filter Toolkit
- Advanced Signal Processing Toolkit
- Database Connectivity Toolkit
- DataFinder Toolkit
- Desktop Execution Trace Toolkit
- Digital Filter Design Toolkit
- Internet Toolkit
- PID and Fuzzy Logic Toolkit
- Report Generation Toolkit
- Simulation Interface Toolkit
- Unit Test Framework Toolkit
- VI Analyzer Toolkit

LabVIEW Modules
- Control Design Simulation Module
- Datalogging and Supervisory Control Module
- FPGA Module
- MathScript Module
- Mobile Module
- Real-Time Module
- Statechart Module
- Touch Panel Module
- Vision Development Module

Other LabVIEW Related Software
- LabVIEW SignalExpress
- Microprocessor SDK
- NI Device Drivers
- NI Motion Assistant
- Sound and Vibration Measurement Suite

1.9 COMPETING PROGRAMS

Agilent VEE is similar to LabVIEW. Like LabVIEW it uses a graphical programming language for instrument control and data acquisition. Through a connection between it and MATLAB, it will also do complicated computations.

Math Works, Inc.'s Simulink graphical programming language is similar to LabVIEW's graphical G programming language.

Computing languages such as C, C#, C++, Delphi, Excel, and Visual Basic can accept data from NI sensing interface devices and those of other companies.

Programmable Logic Controllers (PLCs) will accept data, do simple computations, and control equipment. Where rugged equipment is desired, circuits are not likely to change often, and computations are not complicated, PLCs can compete with LabVIEW.

2.0 LabVIEW TRAINING AND TUTORIALS

There is a great deal of educational material available on LabVIEW.

National Instruments offers free information about LabVIEW in online recorded tutorials, online live tutorials, and live sales seminars. Plus, they have a helpful sales staff. They also offer for-a-fee live classroom tutorials (at National Instruments locations or your site) and online tutorials. See http://www.ni.com/academic/labview_training/ and http://www.ni.com/training/labview.htm for information on National Instruments LabVIEW training. Generally, LabVIEW training costs about $600/day tuition for live classroom and online training.

For-a-fee, LabVIEW tutorials are also available from other companies.

There are many free online LabVIEW tutorials written by people outside of National Instruments. Usually these were written by university faculty who were originally writing tutorials for their students. Some of these are:

1) LabVIEW Tutorials, Written by James R. Drummond of the University of Toronto, http://www.upscale.utoronto.ca/GeneralInterest/LabView.html.

2) How to Create a System Simulation using LabVIEW 8.2, written by Eric Shaffer, Michael Kleinigger, and Kevin Craig, Rensselaer Polytechnic Institute, http://www.eng.mu.edu/~craigk/tutorials/LabVIEW%20SimulationTutorial.pdf.

3) Getting Started with LabVIEW, written by Ed Doering of the Rose-Hulman Institute of Technology, http://cnx.org/content/m14764/latest/.

3.0 DAQS

DAQ is short for **D**ata **Acq**uisition. DAQ devices allow LabVIEW to acquire data from real-world sensors. Many DAQ devices are multifunctional. Besides acquiring data from analog and digital signals they are capable of producing digital and analog outputs. These outputs might control instrumentation, relays, programmable logic controllers, or drives.

A variety of National Instrument DAQs are available, including:

1) Modules that are plugged into a CompactDAQ, CompactDAQ Ethernet, or CompactRIO chassis.

2) PCI and PCI Express modules (cards) that are plugged into a personal computer's motherboard.

3) Modules that are plugged into a PXI chassis.

4) DAQs that are plugged directly into a computer's USB port.

5) Modules that are plugged into a SCXI chassis.

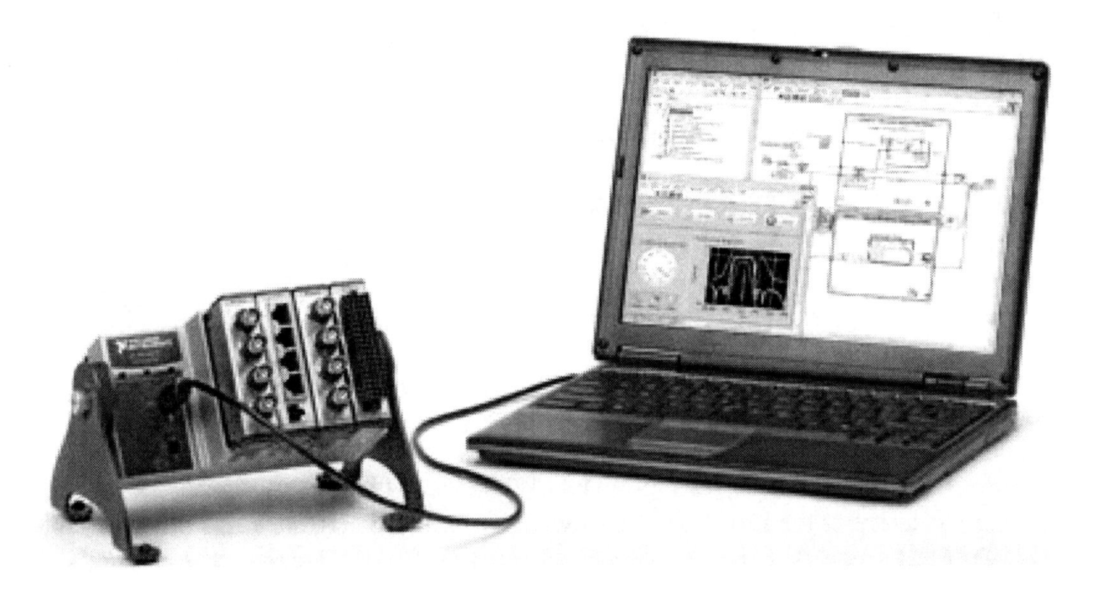

Figure 3-0-1 Laptop running LabVIEW software connected via a USB cable to a NI CompactDAQ chassis with four I/O modules. Photo is courtesy of National Instruments.

Using special software drivers NI DAQs can also convert sensor outputs to formats that can be understood by computers that are running other software. NI DAQs can export signals to computers running software such as, "LabVIEW SignalExpress", "LabWindows/CVI", "Measurement Studio", "ANSI C", "C#", and "Visual Basic.NET".

As with all NI systems and equipment, there is a great deal of DAQ device and driver information available through NI websites. Also, National Instruments offers training on data acquisition through its basic LabVIEW classes.

3.1 DAQ INPUTS

A "NI-DAQ" device input may accept analog, counter, and digital sensor signals. Through various sensors, the signals may represent voltage, temperature, strain, current, resistance, frequency, position, acceleration, humidity, light intensity, sound, force, pressure, fluid flow, pH, switch position, or encoder position. NI has many different types of DAQ devices, from simple one-input types to multifunctional types. Figure 3-1-1 shows the sensor to personal computer signal data flow that goes through a DAQ to LabVIEW.

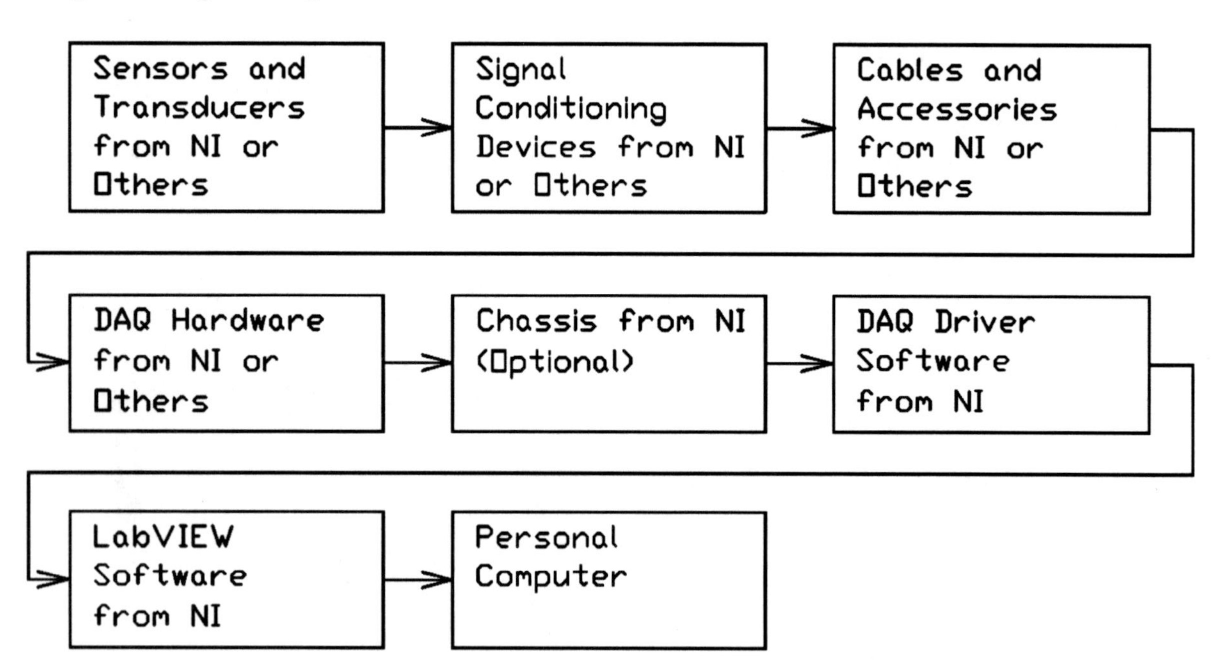

Figure 3-1-1 Flow chart showing the equipment and software that LabVIEW uses to receive real-world data.

National Instruments sells sensors and DAQ devices to convert signals to raw data, and driver software to interpret the DAQ devices' raw data outputs. There are some other manufacturers that make DAQ devices that are compatible with LabVIEW. There are many other sensor manufacturers, such as Honeywell, that manufacture sensors that can be incorporated into a National Instruments system. Through appropriate driver software and DAQ devices LabVIEW can connect to 6000+ instruments from over 250 vendors.

3.2 DAQ OUTPUTS

Some "NI-DAQ" devices are capable of producing digital and/or analog outputs. Figure 3-2-1 shows the personal computer output to controlled equipment data flow that goes through LabVIEW and a DAQ.

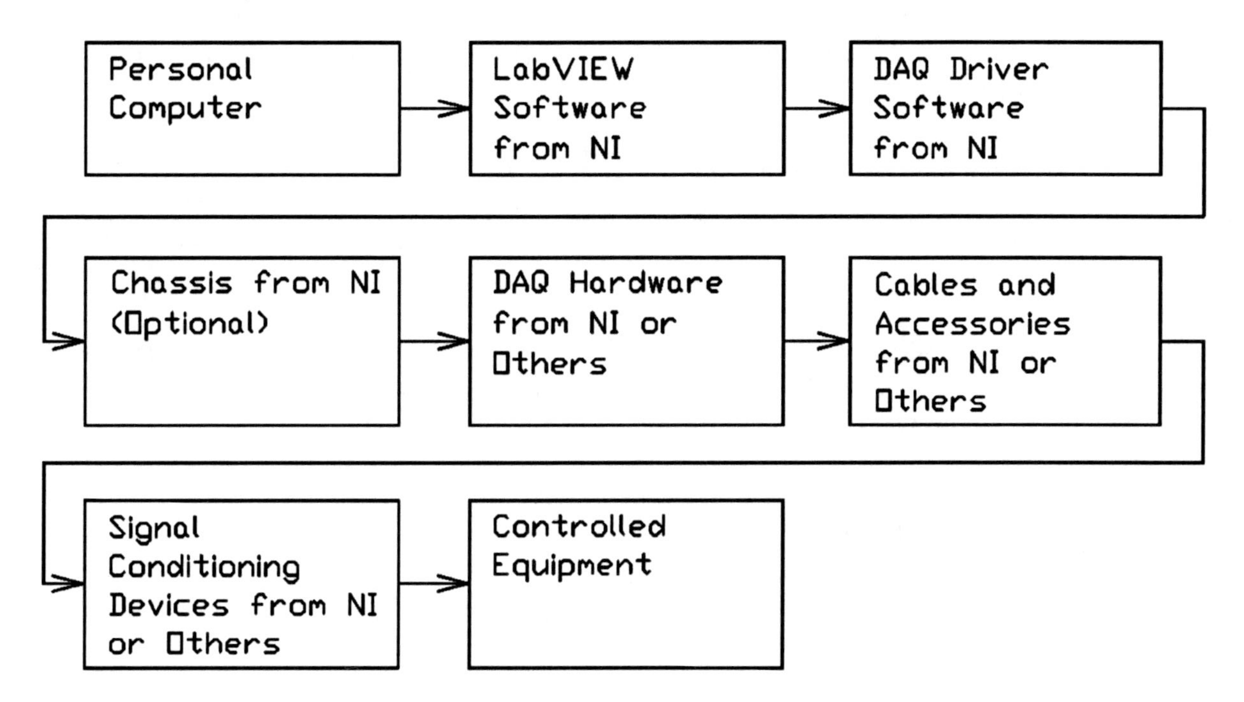

Figure 3-2-1 Flow chart showing the equipment and software that LabVIEW uses to control real-world equipment.

LabVIEW is usually run on a personal computer. The advantages of using a personal computer to control equipment are the software flexibility and the graphical displays. However, one should not forget the reliability of personal computers when considering using one for equipment control. A computer locking up every so often in an office is usually not a serious problem. A personal computer that locks up while controlling a critical process may cause serious problems.

When using LabVIEW with a personal computer, it is better to use DAQ outputs to just control instrumentation and non-critical processes. However, if LabVIEW is run on a NI Single-Board RIO, a solid-state device that does not use a hard drive or Microsoft Windows, it is reasonable to use LabVIEW to control critical processes.

3.3 OBTAINING A DAQ

The most direct way of obtaining a DAQ that can be used with LabVIEW is to purchase it from National Instruments. There are a great number of DAQs available. A LabVIEW sales representative could help in the selection. The least expensive multifunction NI-DAQ, the NI USB-6008, has a list price of $169. The next better quality multifunction NI DAQ, the NI USB-6009, has a list price of $279. Multifunction NI-DAQs that are designed for industrial use have a median list price of about $2,000.

Students and educators can purchase a NI USB-6008 or NI USB-6009 Kit for $169 or $279 respectively. These kits include LabVIEW Student Edition for Windows. Also, the myDAQ, a special DAQ designed for educational use, is available to students and educators for $175.

There are second hand NI-DAQs and off-brand DAQs. However, the buyer should be wary; some of the second-hand NI-DAQs and off-brand DAQs are not totally compatible with newer LabVIEW software.

3.4 DAQ DRIVER SOFTWARE

For LabVIEW to interface with a DAQ, appropriate driver software must be used. Details on installing NI-DAQmx drivers can be found in Section 1.6.

NI calls its older DAQ driver software “Traditional NI-DAQ” and newer “NI-DAQmx”. The two are similar. However, to convert a VI written for a “Traditional NI-DAQ” to that for a “NI-DAQmx”, NI suggests that the VI be rewritten.

This book only considers the “NI-DAQmx” driver software.

4.0 EXAMPLE VIs

The example programs in Sections 4.1 to 4.3 are "data-in/data-out" VIs where actual DAQs (data acquisition devices) are not used. However, in each "data-in/data-out" example, the data that is input from a "Front Panel" input block or a "DAQmx Simulated Device" could have come from an actual DAQ device. Likewise, the data that appears in output blocks on the "Front Panel" could have been directed to an actual DAQ to control real devices.

The examples in Section 4.4 demonstrate LabVIEW receiving real-world electrical data and outputting electrical signals via a DAQ.

The order of the example VIs goes roughly from the simplest and most basic to the more complicated and probably less used. Every beginner should do the first couple examples. Later examples can be studied as needed.

VIs can be constructed from "Standard VI" or "Express VI" blocks. Both are available to the programmer on all LabVIEW versions. "Standard VI" blocks are more versatile, using all of LabVIEW's abilities. "Old-school" experienced LabVIEW programmers are more likely to prefer them. "Express VI" blocks are combinations of select "Standard VI" blocks. "Express VI" blocks were developed to make LabVIEW programming easier for beginners. Both will be demonstrated in the following examples.

4.1 STEADY-STATE ELECTRICAL CIRCUITS

4.1.1 SIMPLE DC CIRCUIT

Problem:

Solve for the resistances of R1 and R2, the voltage VS, and the power to resistor R2 for the circuit of Figure 4-1-1-1. The following values are given: V1 = 5 volts, V2 = 5 volts, and I = 2 amps.

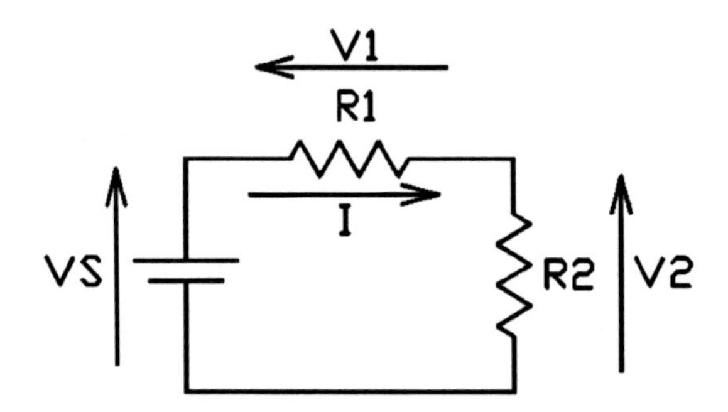

Figure 4-1-1-1 DC supply with a voltage divider.

The equations solving the problem are:

(i) $R1 = V1/I$

(ii) $R2 = V2/I$

(iii) $VS = V1 + V2$

(iv) $P2 = V2 \cdot I$

This problem will be solved two ways, first using G with one VI using ordinary variable blocks and second using G with four "mini" VIs using "Local Variable" blocks.

4.1.1.1 Solution with one VI

1) Start LabVIEW as in Section 1.5. The screen seen in Figure 1-5-1 should appear. Left-click on "Blank VI". The window in Figure 4-1-1-1-1 should appear.

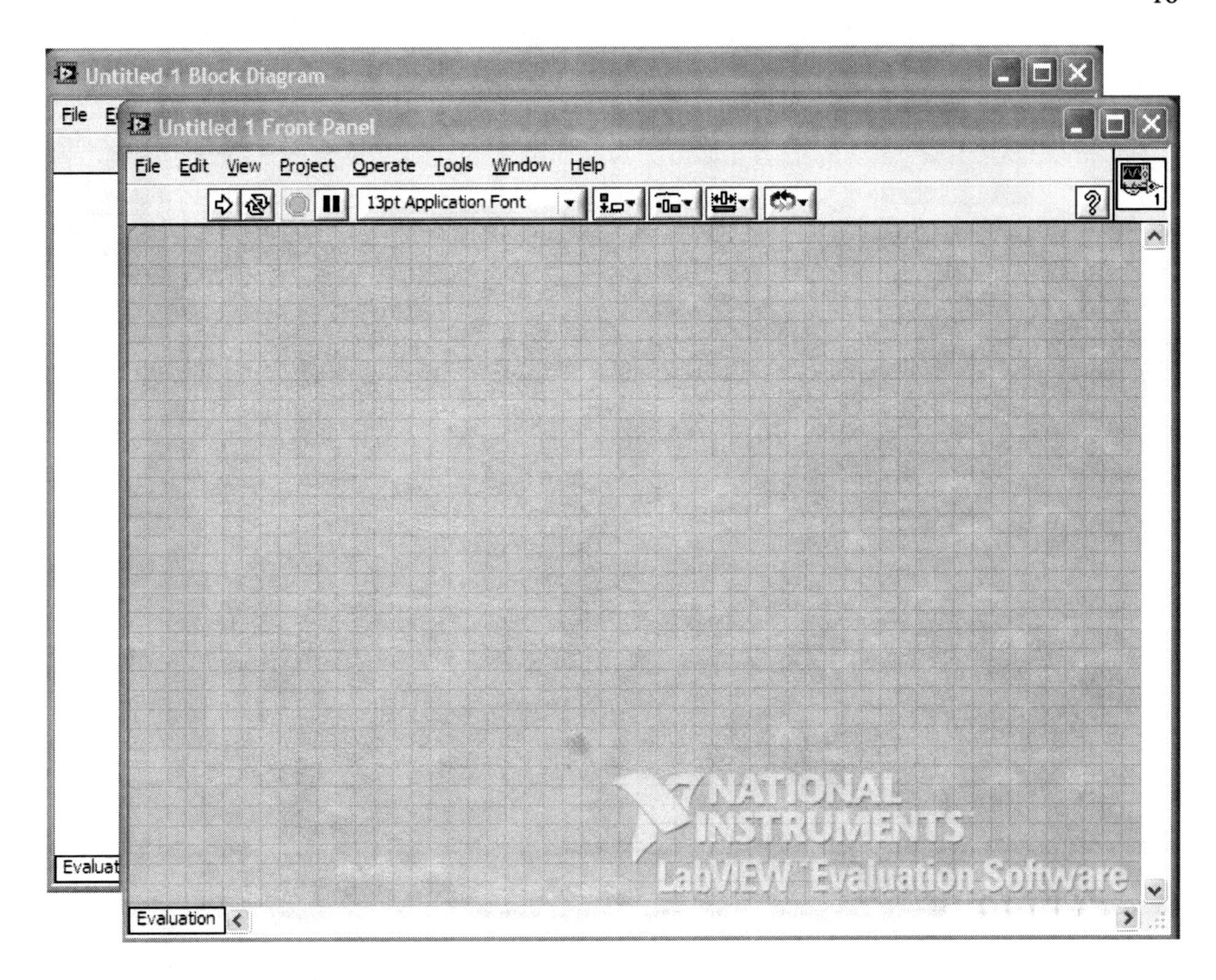

Figure 4-1-1-1-1 Blank “Front Panel” and “Block Diagram”.

2) Put the cursor on the white background window of the “Block Diagram” and left-click. This puts it in the foreground. Alternatively, the “Block Diagram” can be brought up by typing “Ctrl E”.

3) On the “Block Diagram” left-click on “File” and “Save As”. Type “SIMPLE DC CIRCUIT” for the file name and put it in a folder of your choice.

4) First a VI will be written and run for equation (i).

5) Put the title “SIMPLE DC CIRCUIT” on the “Block Diagram”. This is not necessary for running LabVIEW, but it can help the programmer later, when sorting VIs. To place the title, left-click on “View”, ”Tools Palette”<see Section 6.1, page 136, for a description of the “Tools Palette”>, and the “Edit Text” icon (a large A). Then drag the blank text block to where you would like to place text and left-click. Now type in “SIMPLE DC CIRCUIT”. Move the cursor to some position and press “Enter” to set the text in place. The added title is seen in Figure 4-1-1-1-2.

Figure 4-1-1-1-2 "Block Diagram" with title. "Tools Palette" is on the upper right.

6) Open the "Front Panel" by typing "Ctrl E". Put the title "SIMPLE DC CIRCUIT" on the "Front Panel". Use the same method as was used on the "Block Diagram".

7) Left-click on "View", "Controls Palette", Express", and "Numeric Controls". Use the "Numeric Controls" rather than the similar in appearance "Numeric Indicators" that appear just beneath it. Left-click and hold on the "Numeric Control" input block. Drag it to the "Front Panel" and release to set it in place. Do this two more times, so three "Numeric Control" input blocks are set in place. This is shown in Figure 4-1-1-1-3.

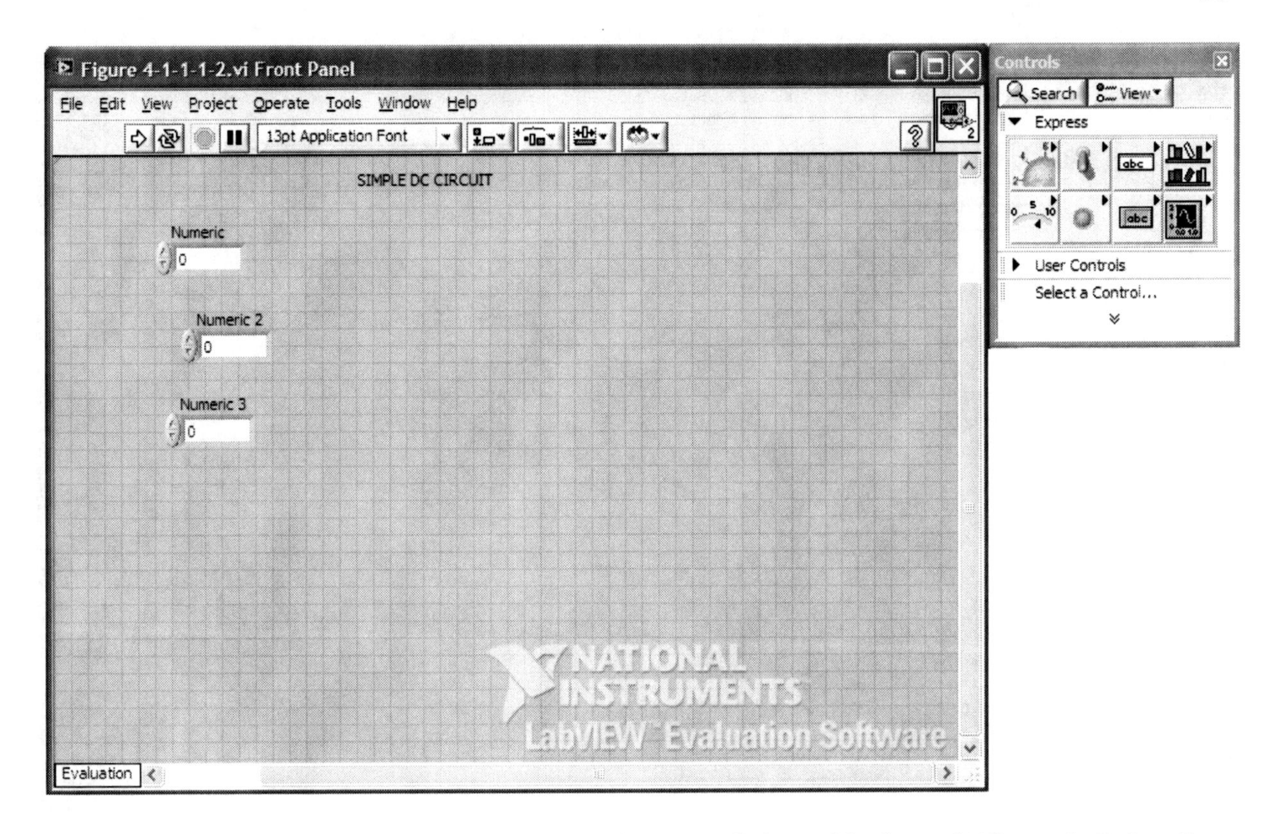

Figure 4-1-1-1-3 "Front Panel" with "Numeric Control" input blocks and "Controls Palette".

8) Overwrite the labels of the input blocks, "Numeric", "Numeric2", and "Numeric3" with "V1", "V2", and "I". To do this left-click on "View", "Tools Palette", and "Edit Text". Then put the cursor on each input block's label, left-click and delete/backspace the old label and put on the new label. Put the cursor someplace else on the "Front Panel" and left-click to set the new label in place.

9) Type "Ctrl E" to switch back to the "Block Diagram". This is shown in Figure 4-1-1-1-4. Notice that data input blocks for V1, V2, and I are shown.

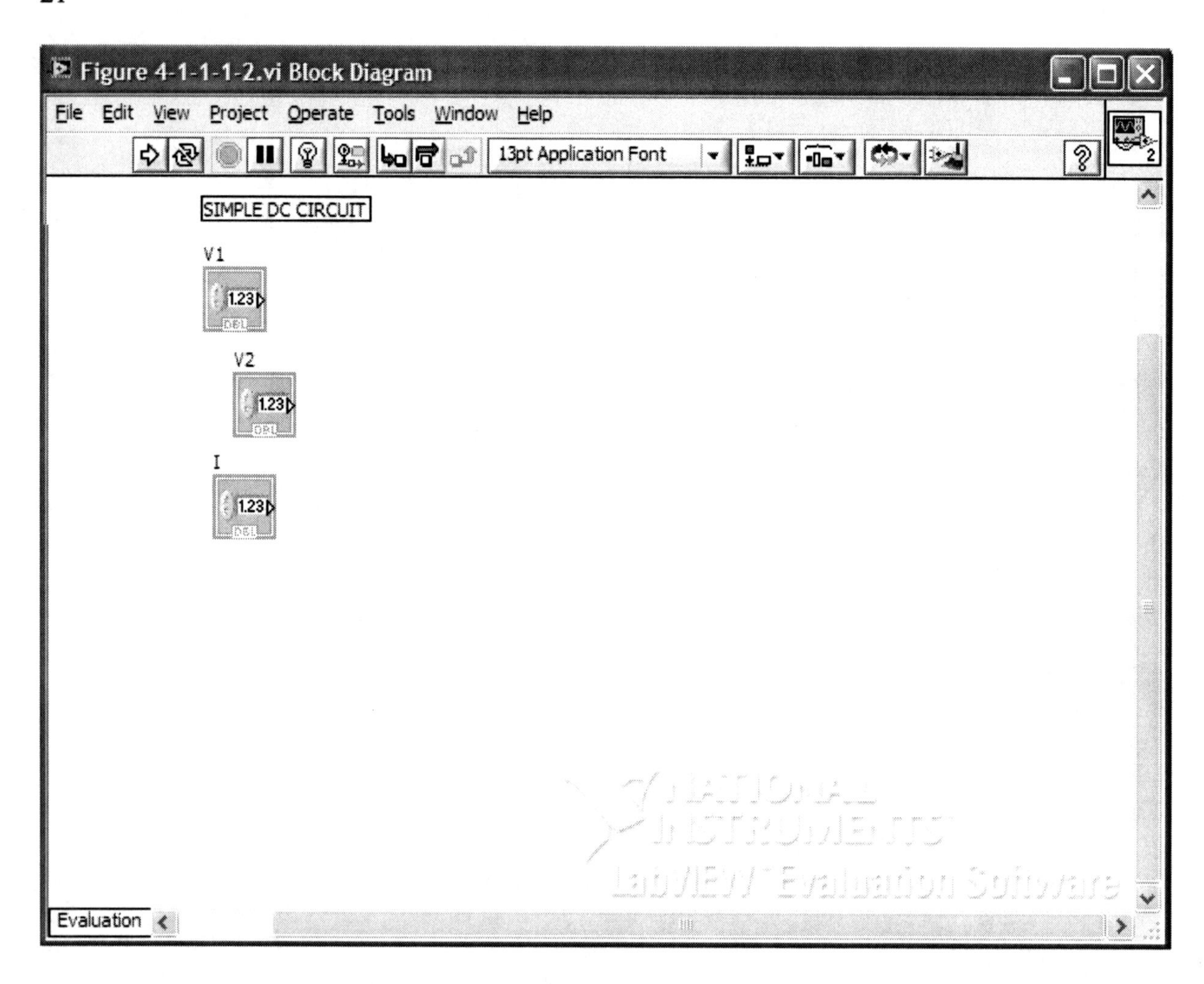

Figure 4-1-1-1-4 "Block Diagram" with V1, V2, and I on it.

10) First, a VI will be written for R1. Later a VI will be written for R2 and P2.

11) On the "Block Diagram", left-click on "View", "Functions Palette", "Programming", "Numeric". Left-click, hold down the mouse button and drag the "Divide" function block to the "Block Diagram" Release the left mouse button to set it in place. This is shown in Figure 4-1-1-1-5.

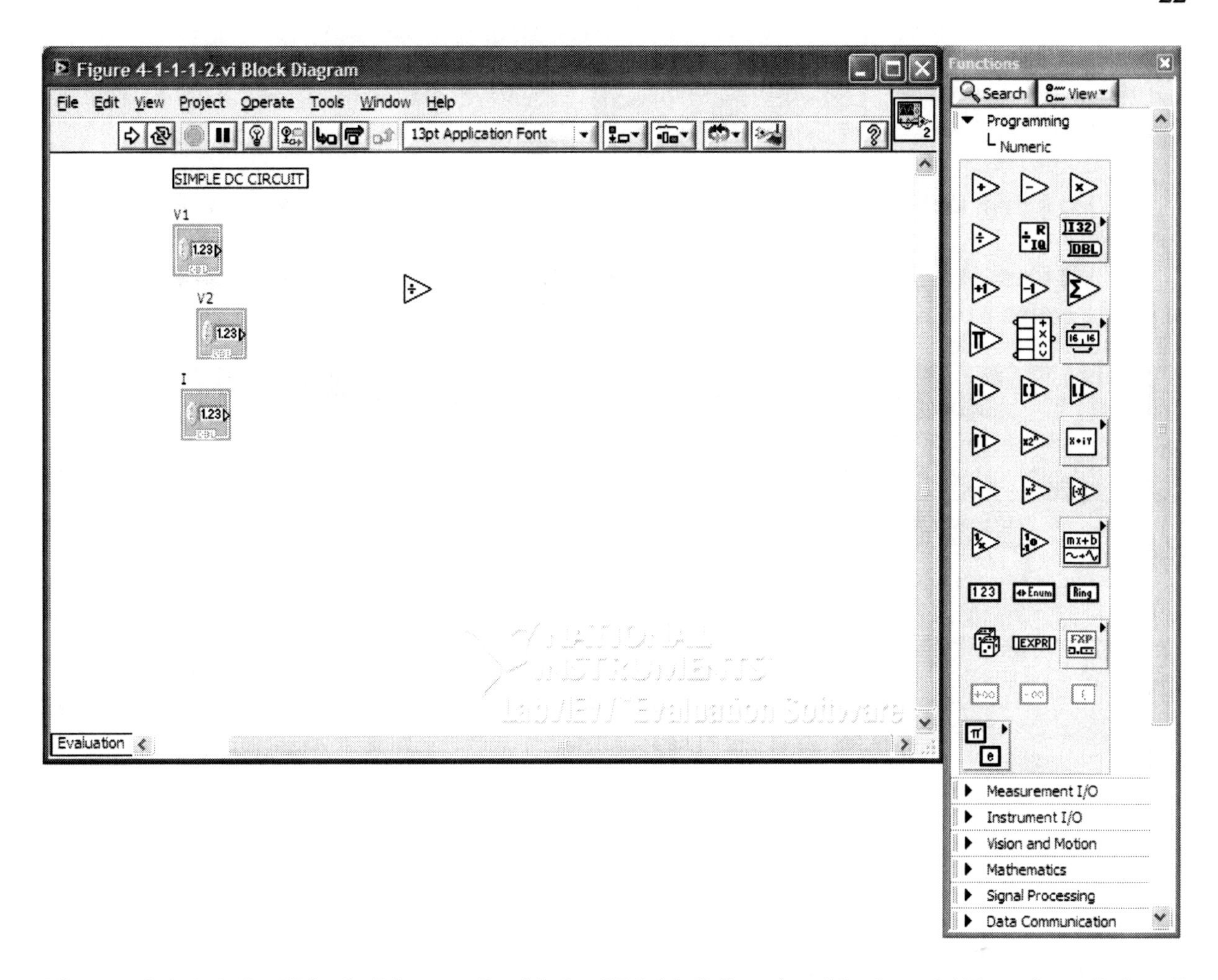

Figure 4-1-1-1-5 "Block Diagram" with its "Divide" function block and "Functions Palette" with its "Programming" "Numeric" function blocks.

12) Now the data value from the V1 input block needs to flow to the "Divide" function block where it will be divided by the data value from the R1 input block. To make the dataflow wire from V1 left-click on "View", "Tools Palette", and the "Connect Wire" icon (a wire spool). The cursor becomes a wire spool icon. It can be dragged by simply moving the mouse. Drag the wire spool icon to the right-side middle of the V1 input block. It should start flashing and a red connection dot should be visible. Left-click to set the dataflow wire. Drag the other end of the dataflow wire to the upper left-side of the "Divide" function block. This icon should also start flashing and show a red connection dot. Left-click again to connect the dataflow wire. The properly connected dataflow wire is solid orange, a black dashed line would indicate a broken dataflow wire.

13) In the same way connect a dataflow wire from the I input block to the lower left-side of the "Divide" function block.

14) Return to the "Front Panel". On this left-click "View", "Controls Palette", "Express", and "Numeric Indicators". Left-click the "Numeric Indicator" output block and drag it onto the "Front Panel". Change its label from "Numeric" to "R1".

15) Return to the "Block Diagram". There should now be a R1 output block on it.

16) Connect a dataflow wire from "Divide" function block right-side to the middle left-side of the R1 output block. This is shown in Figure 4-1-1-1-6.

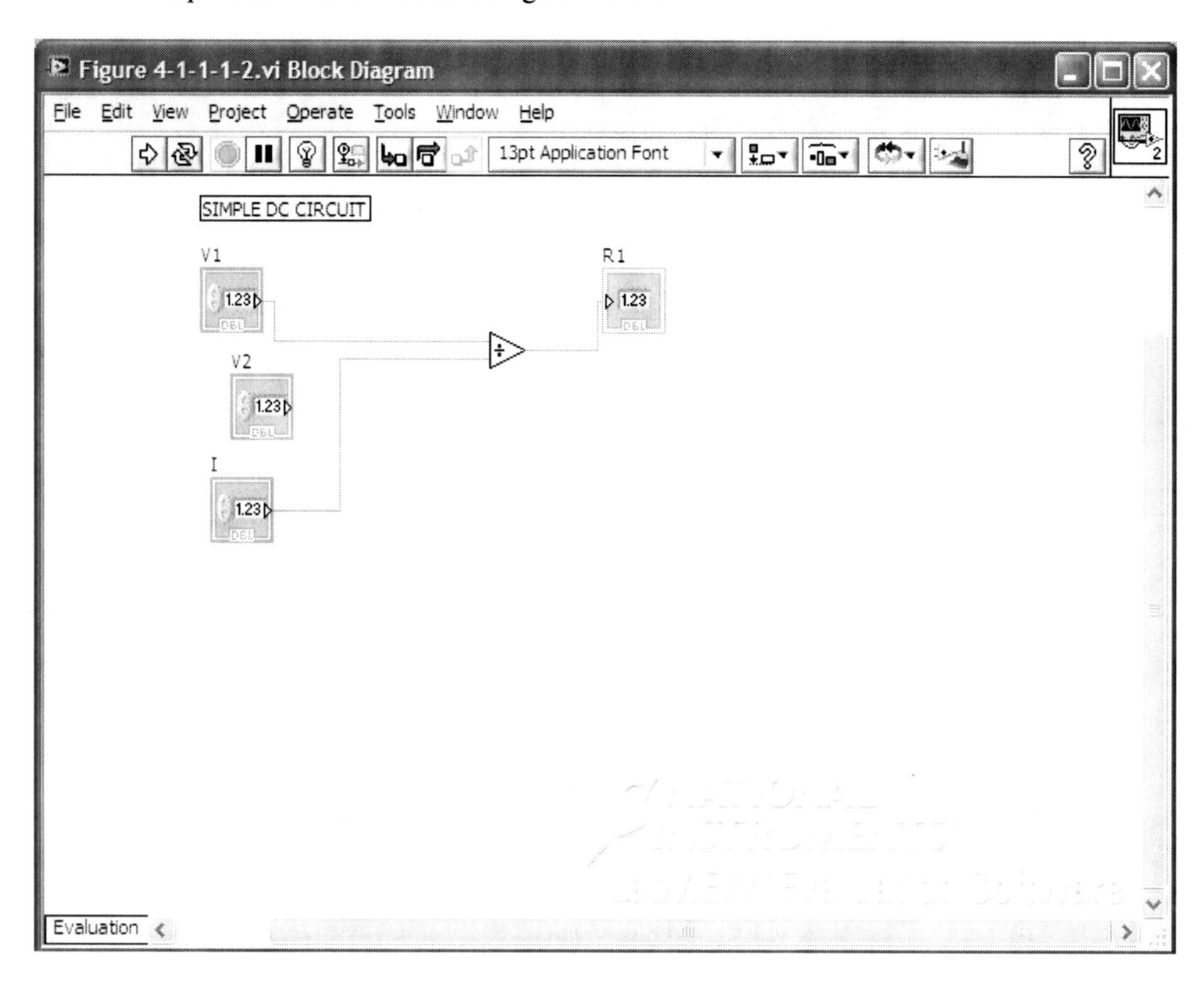

Figure 4-1-1-1-6 Completed "Block Diagram" for calculating R1.

17) Return to the "Front Panel". Left-click on "View", "Tools Palette", and "Operate Value". This turns the cursor into a pointing hand icon.

18) Put the pointing hand icon on the space in the V1 input block and left-click highlight the existing 0 value. Type in the value 5 and <Enter>. In the same way put the pointing hand on the space in the I input block and type in the value 2.

19) On the "Block Diagram" and "Front Panel" under the "Edit" and "View" menu headings there are right pointing arrows. Left-click on one of those arrows to run the VI. The result is in Figure 4-1-1-1-7.

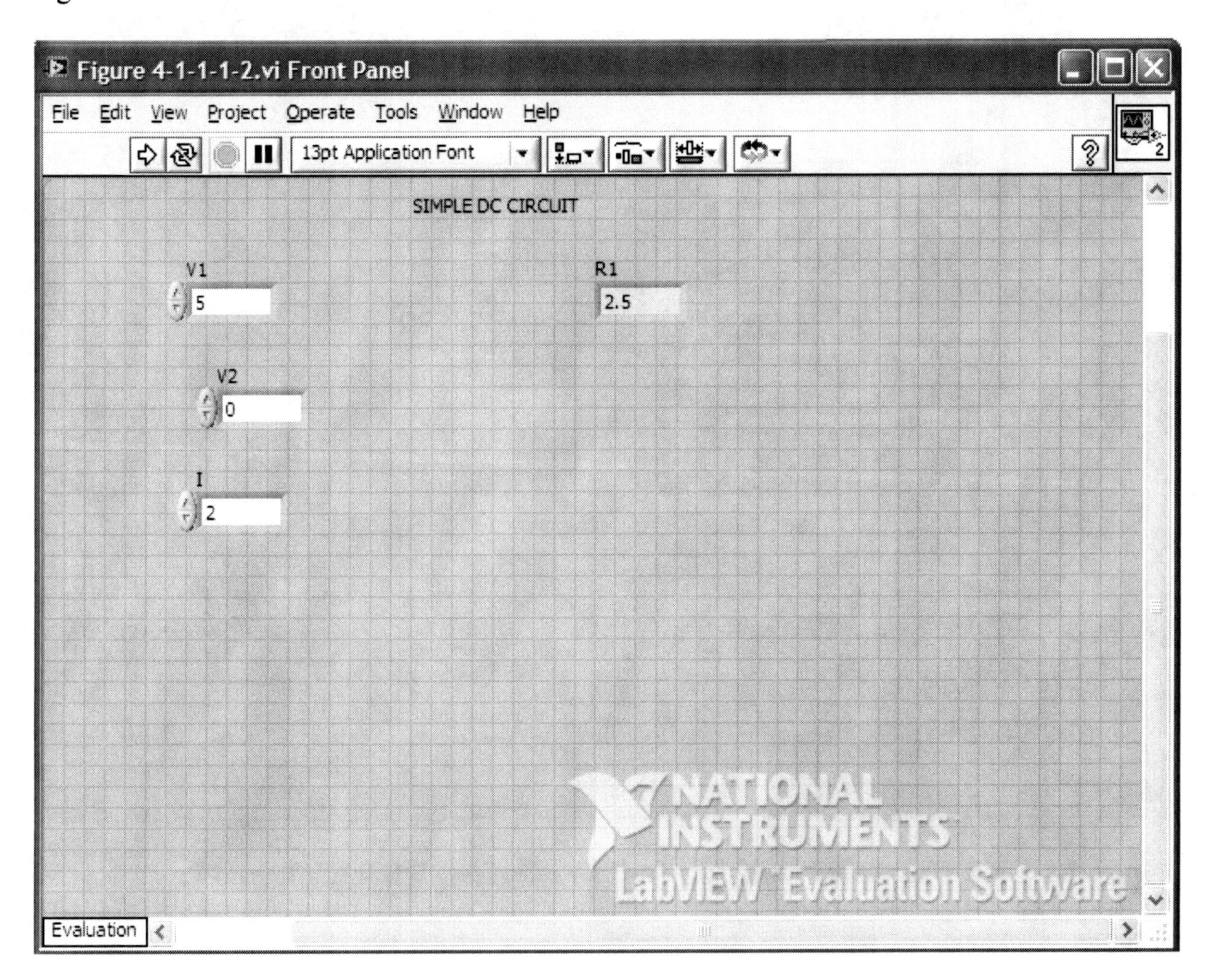

Figure 4-1-1-1-7 Completed and run "Front Panel" for calculating R1.

20) R2, VS, and P2 are solved by inserting appropriate "Add", "Divide", and "Multiply" function blocks and dataflow wiring into the "Block Diagram" and appropriate output blocks in the "Front Panel". The "Block Diagram is in Figure 4-1-1-1-8. The "Front Panel" is in Figure 4-1-1-1-9.

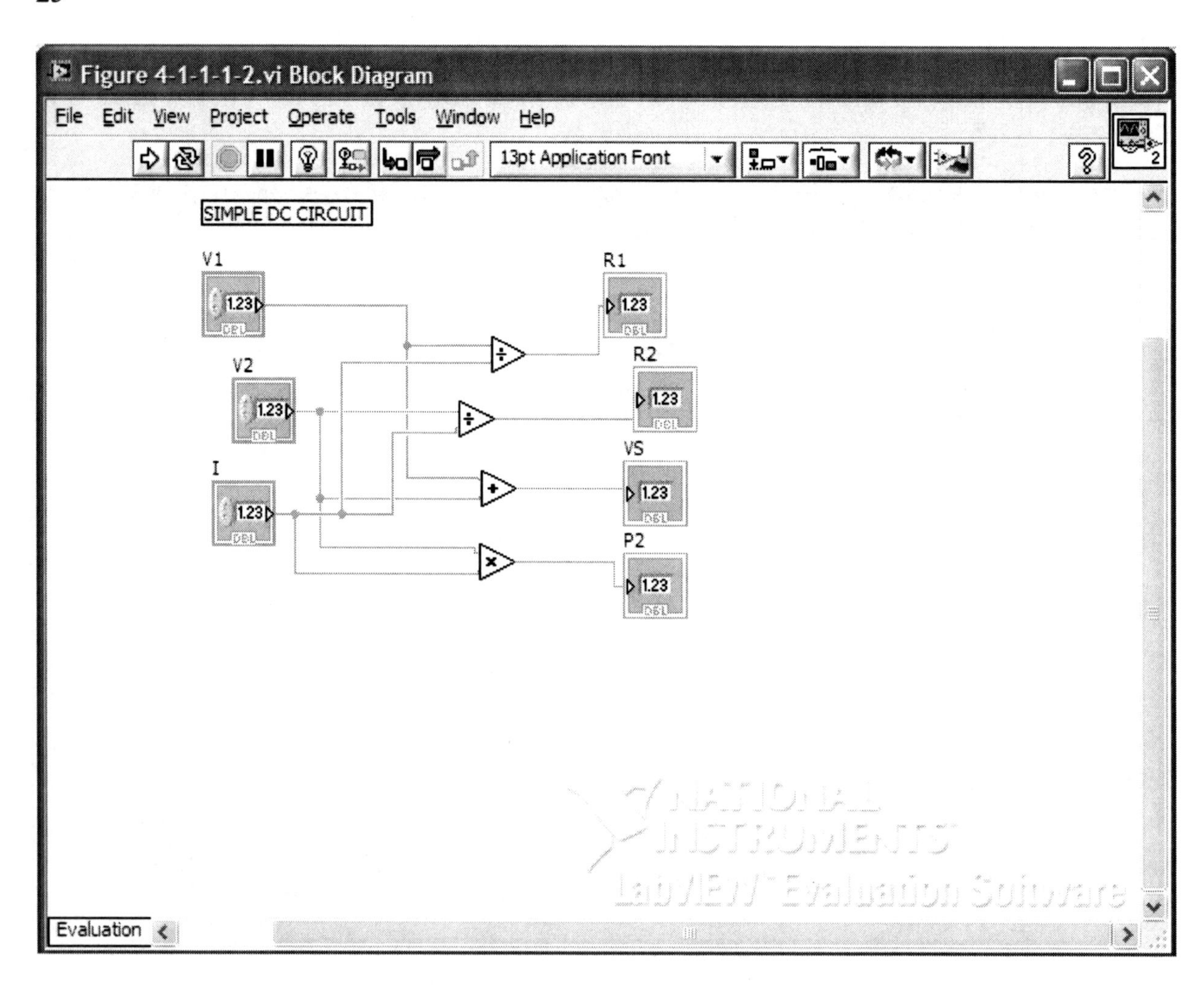

Figure 4-1-1-1-8 Completed "Block Diagram" for calculating R1, R2, VS, and P2.

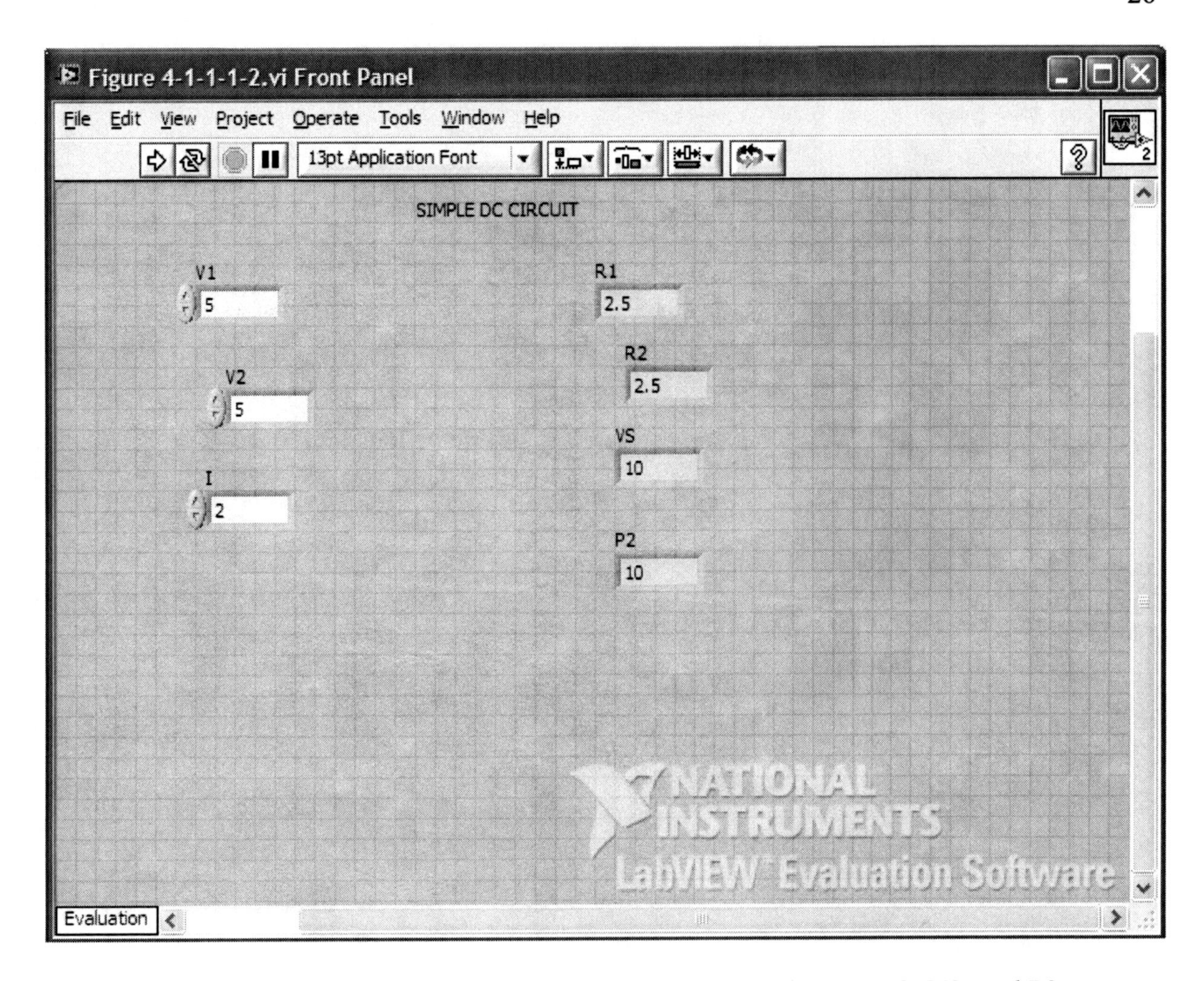

Figure 4-1-1-1-9 Completed and run "Front Panel" for calculating R1, R2, VS, and P2.

21) LabVIEW has editing features to automatically neaten and align "Block Diagram" drawings and to align the "Front Panel" blocks.

22) To neaten the "Block Diagram", left-click its "Edit" and "Clean Up Diagram". The result can be seen in Figure 4-1-1-1-10.

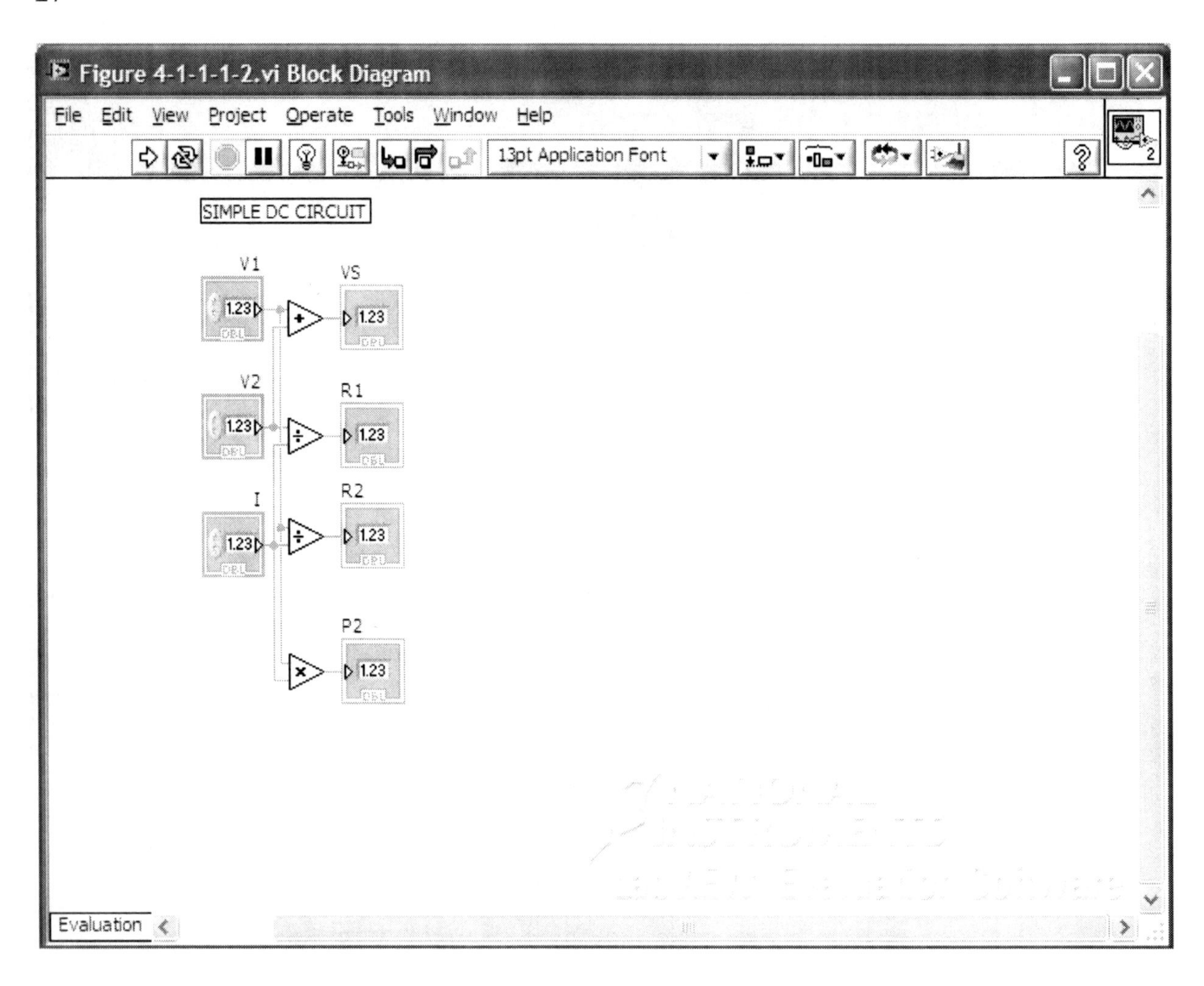

Figure 4-1-1-1-10 Cleaned up "Block Diagram" for calculating R1, R2, VS, and P2 after clean up.

23) To neaten the "Front Panel", left-click its "View", "Tools Palette", "Position/Size/Select" icon. Then position the cursor above the upper left of the three input blocks, V1, V2, and I. Hold down the left mouse button and pull a dashed line rectangle over the three inputs. Release the left mouse button to cause the three to be highlighted. Now left-click on "Edit" and "Align Items". This aligns the three horizontally. The same is done with the four output blocks, R1, R2, VS, and P2. See the result in Figure 4-1-1-1-11.

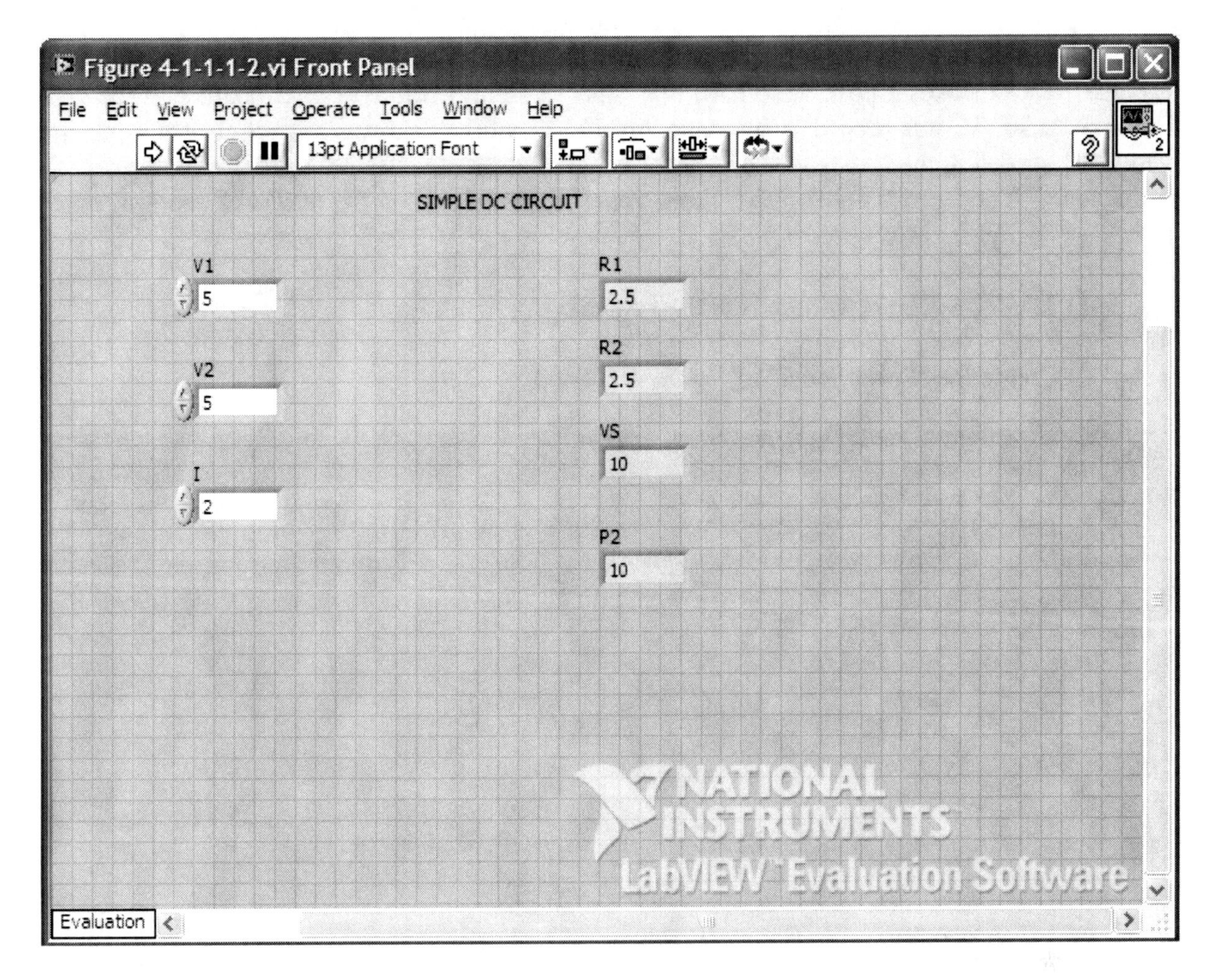

Figure 4-1-1-1-11 Completed and run "Front Panel" for calculating R1, R2, VS, and P2 after alignment.

4.1.1.2 Solution with "Local Variable" Blocks:

"Local Variable" "read" blocks make it possible to have one input block on the "Front Panel" send data to two or more "read" blocks on the "Block Diagram".

Here "Local Variable" "read" blocks will be used to send the data from one "Front Panel" input block to four "mini" VIs. The advantage of these is that they simplify the VIs for the programmer. The "Block Diagram" VI shown in Figure 4-1-1-1-10 is difficult to check relative to that of the four separate "mini" VIs of Figure 4-1-1-2-2.

However, there can be a timing problem with "Local Variable" blocks. The use of "Local Variable" blocks makes it possible to have one "mini" VI finish before or after another. If the data from one "mini" VI was needed by a second "mini" VI, but the data was not yet created when the second VI needed it, the results would be incorrect. This example does not have this problem, each of the "mini" VIs depends only on the "Front Panel" data and not on the results of another "mini" VI.

1) The VI will be rewritten from the beginning.

2) Create the "Front Panel" of Figure 4-1-1-2-1.

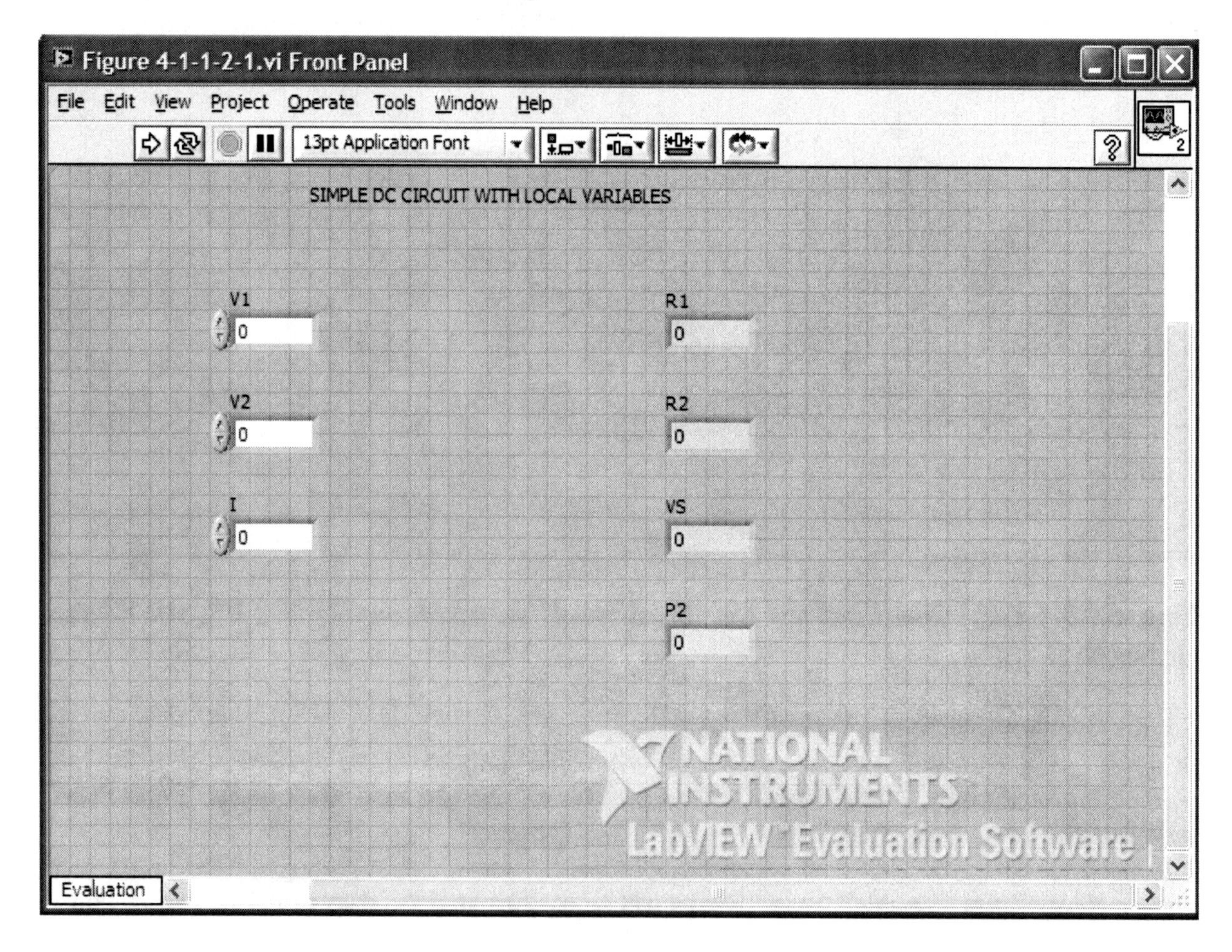

Figure 4-1-1-2-1 "Front Panel" with ordinary variables.

3) Switch to the "Block Diagram".

4) On the "Block Diagram" right-click on the V1 input block. Then left-click on "Create" and "Local Variable". This will make a "write" "Local Variable" block for V1 in the "Block Diagram". Right-click on this and then left-click on "Change to Read". This makes the block a "read" "Local Variable" that can be used as an input to a function block.

5) Create "Local Variable" "read" blocks from theV1, V2, and I input blocks and connect them with math function blocks to create the four VIs shown in Figure 4-1-1-2-2.

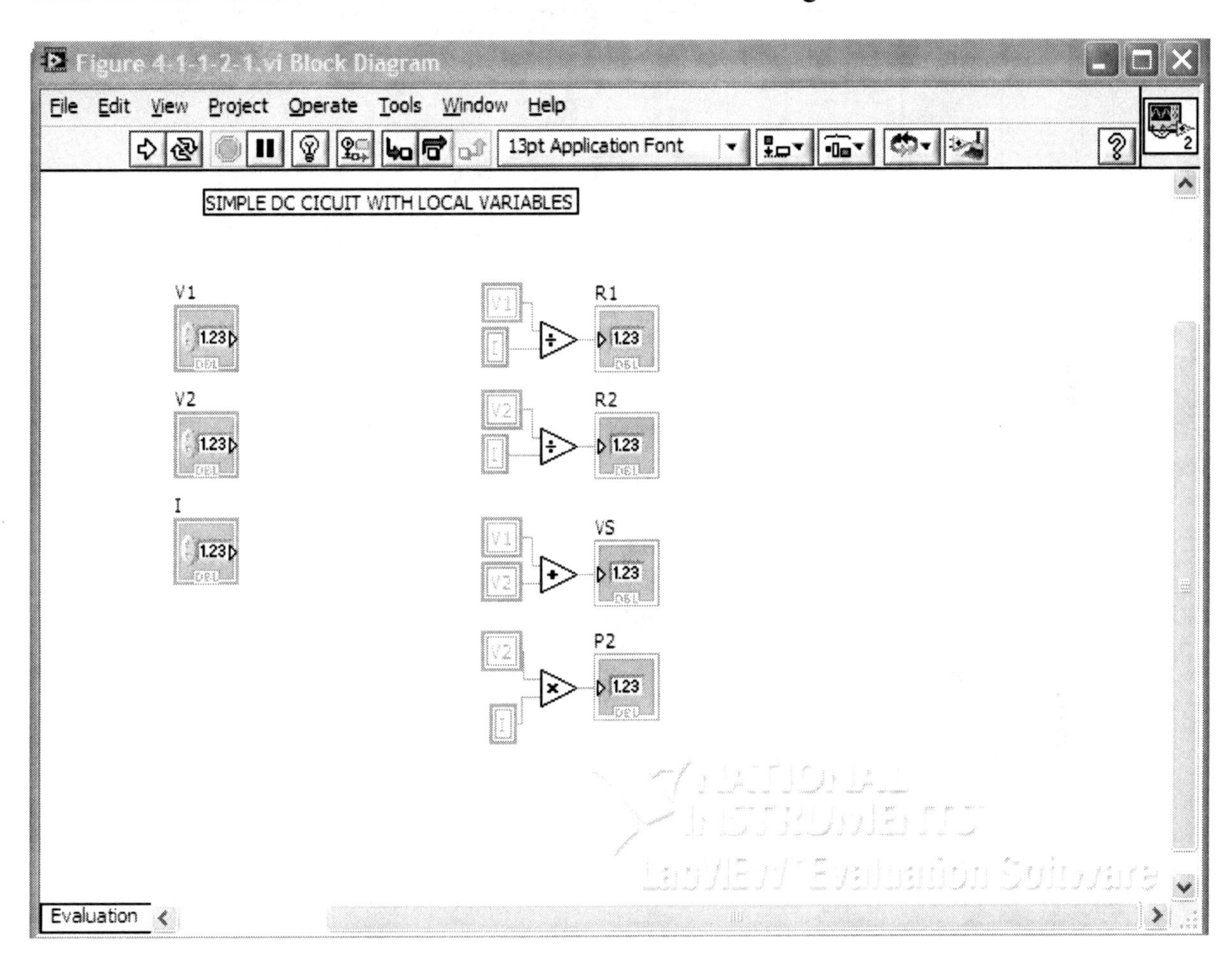

Figure 4-1-1-2-2 "Block Diagram" with ordinary variables, added "Local Variable" "read" blocks, and math function blocks.

6) Running this VI produces the same results as seen in Figure 4-1-1-1-11.

4.1.2 DC MESH CIRCUIT

Problem:

Solve for the currents I1, I2, and I3 in the circuit of Figure 4-1-2-1. The following values are given: VS = 10 volts, R1 = 1 Ω, R2 = 2 Ω, R3 = 3 Ω, R4 = 4 Ω, R5 = 5 Ω, and R6 = 6 Ω.

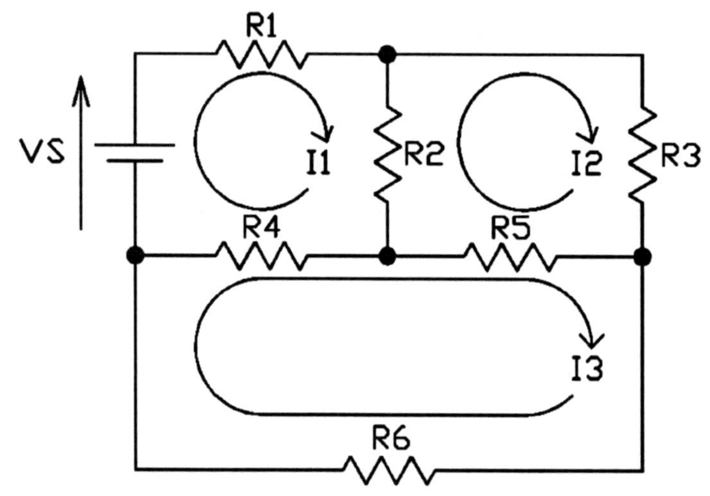

Figure 4-1-2-1 DC mesh circuit.

The simultaneous equations that describe Figure 4-1-2-1 are:

$$VS = I1 \cdot (R1 + R2 + R4) + I2 \cdot (-R2) + I3 \cdot (-R4)$$
$$0 = I1 \cdot (-R2) + I2 \cdot (R2 + R3 + R5) + I3 \cdot (-R5)$$
$$0 = I1 \cdot (-R4) + I2 \cdot (-R5) + I3 \cdot (R4 + R5 + R6)$$

In matrix form the equations would be written:

$$\begin{pmatrix} VS \\ 0 \\ 0 \end{pmatrix} = \begin{pmatrix} (R1+R2+R4) & -R2 & -R4 \\ -R2 & (R2+R3+R5) & -R5 \\ -R4 & -R5 & (R4+R5+R6) \end{pmatrix} \bullet \begin{pmatrix} I1 \\ I2 \\ I3 \end{pmatrix}$$

LabVIEW can solve for the currents using determinants and Cramer's rule or using matrix inversion and multiplication. Here the currents will be found using LabVIEW's matrix inversion function block to create the inverse resistance matrix and then multiplying that by the voltage matrix with the matrix multiplication function block.

$$\begin{pmatrix} I1 \\ I2 \\ I3 \end{pmatrix} = \begin{pmatrix} (R1+R2+R4) & -R2 & -R4 \\ -R2 & (R2+R3+R5) & -R5 \\ -R4 & -R5 & (R4+R5+R6) \end{pmatrix}^{-1} \bullet \begin{pmatrix} VS \\ 0 \\ 0 \end{pmatrix}$$

LabVIEW's matrix input, output, inversion and multiplication blocks are demonstrated here.

Solution:

1) The resistance matrix with its numbers inserted is:

$$\begin{pmatrix} 7 & -2 & -4 \\ -2 & 10 & -5 \\ -4 & -5 & 15 \end{pmatrix}$$

2) The voltage matrix with its numbers inserted is:

$$\begin{pmatrix} 10 \\ 0 \\ 0 \end{pmatrix}$$

3) Start LabVIEW and create a "Front Panel" with three matrices. To create a matrix left-click on "View", "Controls Palette", "Classic", "Classic Array, Matrix & Cluster". Left-click and drag "Real Matrix ctl" blocks onto the "Front Panel". Create three of these matrices. Label the matrices R, V, and I. This can be seen in Figure 4-1-2-2.

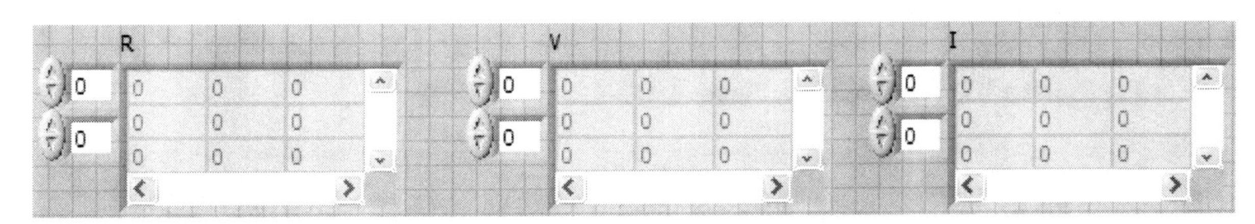

Figure 4-1-2-2 "Front Panel" with matrices before data is entered.

4) Open the “Block Diagram”.

5) Change the I matrix to an output (indicator) block by right-clicking on it and left-clicking on “Change to Indicator”.

6) Insert an “Inverse Matrix” function block between the R and V. To do this, left-click on “View”, “Functions Palette”, “Mathematics”, and “Linear Algebra”. Left-click and drag the “Inverse Matrix” function block onto the “Front Panel”.

7) Insert an “A x B.vi” matrix multiplication function block between the V and I. To do this, left-click on “View”, “Functions Palette”, “Mathematics”, and “Linear Algebra”. Left-click and drag the “A x B.vi” function block onto the “Front Panel”.

8) Connect the dataflow wires between the input, function, and output blocks. The result is shown in Figure 4-1-2-3.

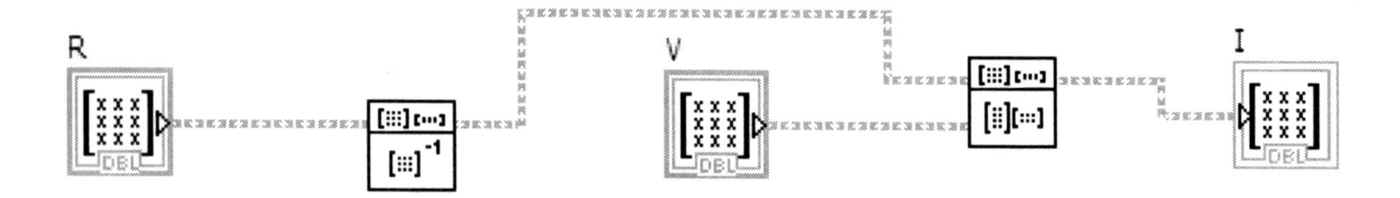

Figure 4-1-2-3 “Block Diagram” with matrices before data is entered.

9) Open the “Front Panel” again.

10) Look at the matrix blocks. On the left of each matrix block are two cell location numbers. The lower is horizontal location, the upper is vertical location. These allow you to move to cells far beyond the nine shown in the grid on the right.

11) Enter the R and V matrix cell values.

12) Run the VI. The result is in Figure 4-1-2-4.

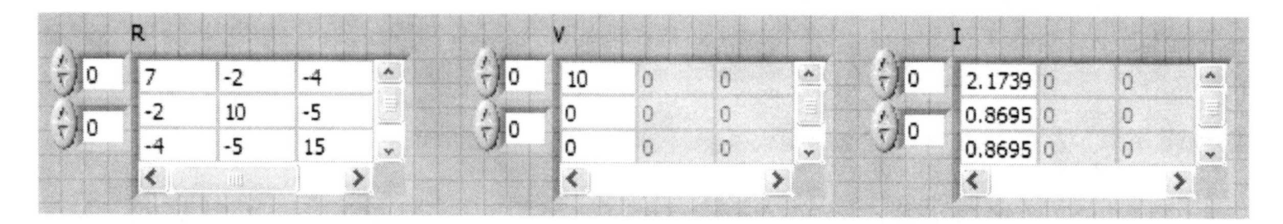

Figure 4-1-2-4 “Front Panel” with matrix input data and its resulting calculated I matrix.

4.1.3 SIMPLE AC PHASOR CIRCUIT

LabVIEW represents complex numbers as single entities rather than two separate numbers. To get a complex number's real and imaginary, or magnitude and angle values conversion functions are used. Angles are in radians.

In this book's text and figures, phasor values will be indicated by larger bold print. For example, the phasor value for VS is **VS**. However, it is not possible to change the font of individual characters in a LabVIEW VI.

Problem:

Determine the values of R1 and XC1 for the circuit of Figure 4-1-3-1. The supply voltage, **VS**, is $10\angle 0^{\circ}$ volts rms and the current is $2.357\angle 45^{\circ}$ amps rms.

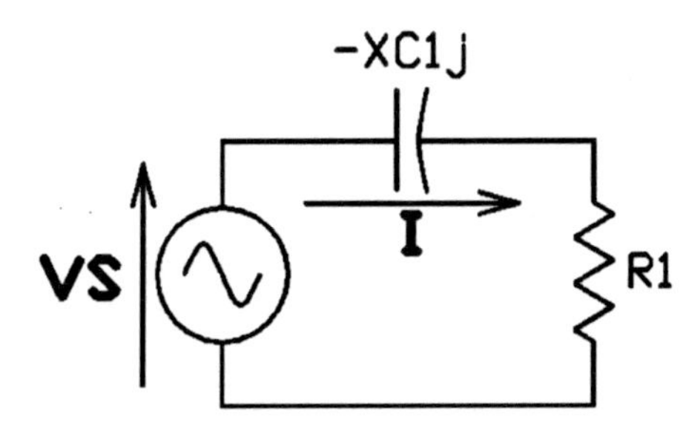

Figure 4-1-3-1 AC supply with a RC voltage divider.

LabVIEW's use of complex number and absolute value functions are demonstrated here.

Solution:

1) The equation for the circuit is **VS** = **I**·(R1 – XC1j)

2) R1 = Real part of (**VS/I**)
XC1 = Imaginary part of (**VS/I**)

3) Write and run the VI of Figure 4-1-3-2. and Figure 4-1-3-3. To get the complex number conversion functions for the "Block Diagram", left-click on "View", "Functions Palette", "Mathematics", "Numeric", and "Complex". From this, drag and set the "Polar to Complex" function block and "Complex to Re/Im" function block <Complex Number Converted to Real and Imaginary Parts> onto the "Block Diagram". To get the absolute value of XC1, left-click on "View", "Functions Palette", "Mathematics", and "Numeric. From this drag the "Absolute Value" function block onto the "Block Diagram".

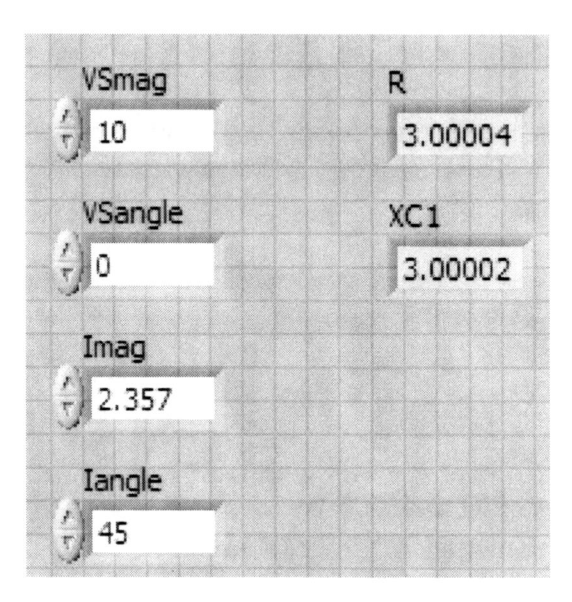

Figure 4-1-3-2 "Front Panel" with input data and solution data.

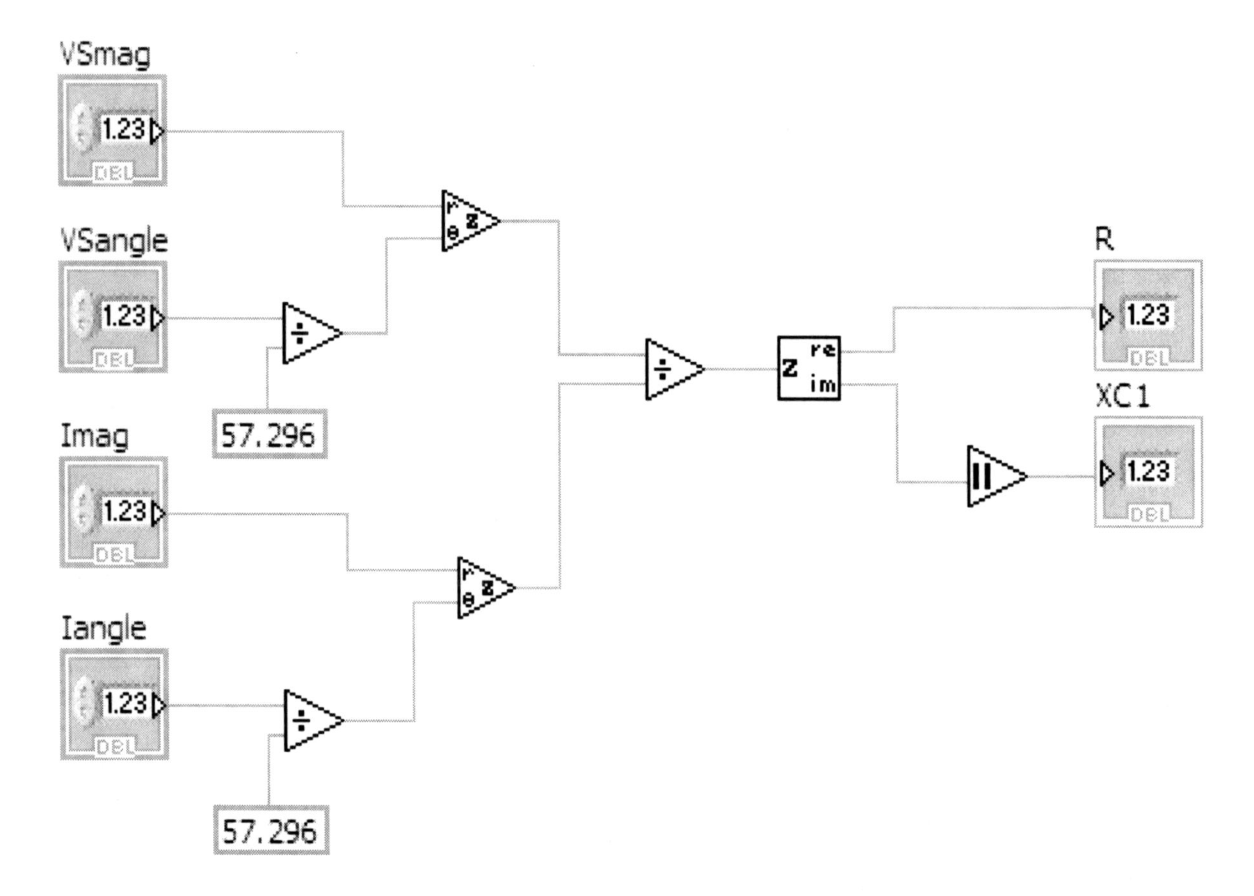

Figure 4-1-3-3 "Block Diagram" using "Polar to Complex", "Complex to Re/Im", and "Absolute Value" function blocks.

4.1.4 AC PHASOR MESH CIRCUIT

Problem:

Solve for the phasor currents **I1**, **I2**, and **I3** in the circuit of Figure 4-1-4-1. The following values are given: VS = 10 volts rms, phase angle of VS = 0, R1 = 1 Ω, R2 = 4 Ω, XL1 = 2 Ω, XL2 = 5 Ω, XC1 = 3 Ω, and XC2 = 6 Ω.

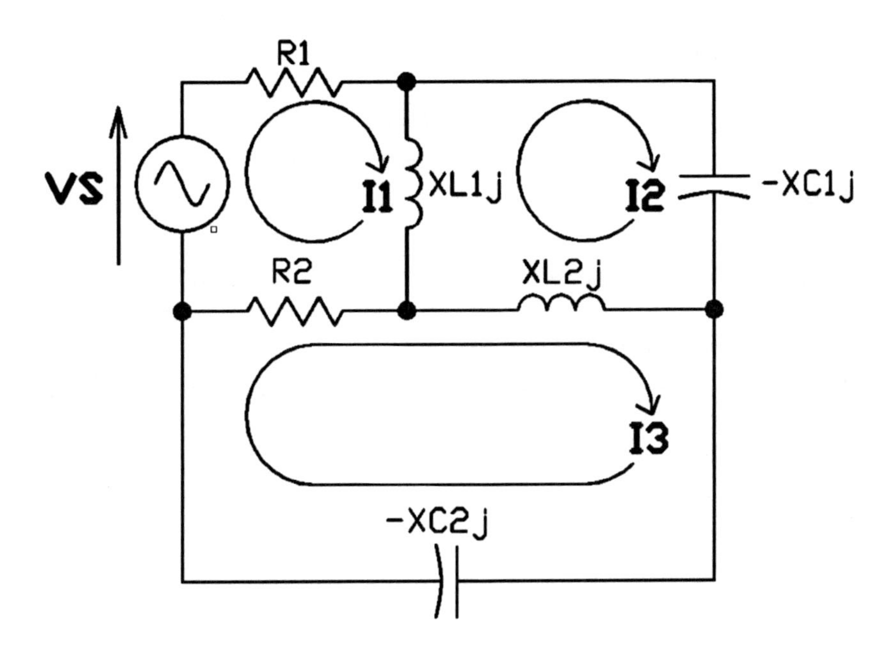

Figure 4-1-4-1 AC mesh circuit.

LabVIEW's complex number matrix and "Negate" function blocks are demonstrated here.

Solution:

1) The equations for this mesh circuit are:

$$\mathbf{VS} = \mathbf{I1}\cdot(R1 + R2 + XL1j) + \mathbf{I2}\cdot(-XL1j) + \mathbf{I3}\cdot(-R2)$$
$$0 = \mathbf{I1}\cdot(-XL1j) + \mathbf{I2}\cdot[(XL1 + XL2 - XC1)j] + \mathbf{I3}\cdot(-XL2j)$$
$$0 = \mathbf{I1}\cdot(-R2) + \mathbf{I2}\cdot(-XL2j) + \mathbf{I3}\cdot[R2 + (XL2 - XC2)j]$$

2) This circuit is solved by the same method used in the analysis of the DC mesh circuit in Section 4.1.2, multiplication of the inverse impedance matrix by the voltage matrix.

3) Place and label the impedance values and source voltage on the "Front Panel".

4) Put complex matrices on the "Front Panel", by left-clicking on "View", "Controls Palette", "Modern", and "Array, Matrix & Cluster". Then left-click, drag, and set four "Complex Matrix.ctl" blocks onto the "Front Panel". Label them "Initial Blank Matrix", "Impedance Matrix", "VS Matrix", and "I Matrix".

5) Go to the "Block Diagram" and change the "Impedance Matrix", "VS Matrix", and "I Matrix" blocks from inputs (controls) to outputs (indicators) by right-clicking on them and left-clicking on "Change to Indicator".

6) On the "Block Diagram" go to the complex matrix element setting function block by left-clicking on "View", "Functions Palette", "Programming", "Array", and "Matrix". Then left-click, drag, and set the "Set Matrix Elements" function block onto the "Block Diagram".

7) Stretch the "Set Matrix Elements" to allow space for nine matrix elements.

8) Connect "Numeric Constant" input blocks to the "Set Matrix Elements" function block to represent row and column locations. Rows are 0, 1, or 2. Columns are 0, 1, or 2. Get the "Numeric Constant" input blocks by left-clicking on "View", "Functions Palette", "Programming", and "Numeric". Then left-click, drag and set "Numeric Constant" input blocks as needed.

9) Use the neaten feature to improve the "Block Diagram" readability. To get it left-click on "Edit" and "Clean Up Diagram".

10) Create the VI that puts R1, R2, XL1, XL2, XC1, XC2, and VS into the complex number equations needed in the impedance and voltage matrices. This VI will require "Add", "Subtract", "Negate", "Re/Im to Complex" function blocks and dataflow wiring. See Figure 4-1-4-3.

11) To get the complex number conversion function block, left-click on "View", "Functions Palette", "Mathematics", "Numeric", and "Complex". From this, drag and set the "Re/Im to Complex" function block onto the "Block Diagram".

12) A "Negate" function block is used to create negatives of positive inputs. To get the "Negate" function block, left-click on "View", "Functions Palette", "Programming", and "Numeric". From this drag and set the "Negate" function block on the "Block Diagram" where appropriate.

13) Insert an "Inverse Matrix" function block. To get the "Inverse Matrix" function block, left-click on "View", "Functions Palette", "Mathematics", and "Linear Algebra". Left-click and drag it onto the "Front Panel".

14) Insert an "A x B.vi" matrix multiplication function block. To get the "A x B.vi" function block, left-click on "View", "Functions Palette", "Mathematics", and "Linear Algebra". Left-click and drag the "A x B.vi" function block onto the "Front Panel".

15) Insert an "I Matrix" output block to receive the result of the matrix multiplication.

16) Use the "Clean Up Diagram" command again.

17) When the VI is run, the "I Matrix" output block will need to be dragged out wider to display the complete answer. To drag it out, left-click on "Position/Size/Select" mode in the "Tools Palette", put the cursor on the right border of the cell to be widened, then left-click and hold down the button and drag the cell's border to the right as needed.

18) The VI is shown in Figure 4-1-4-2 and 4-1-4-3.

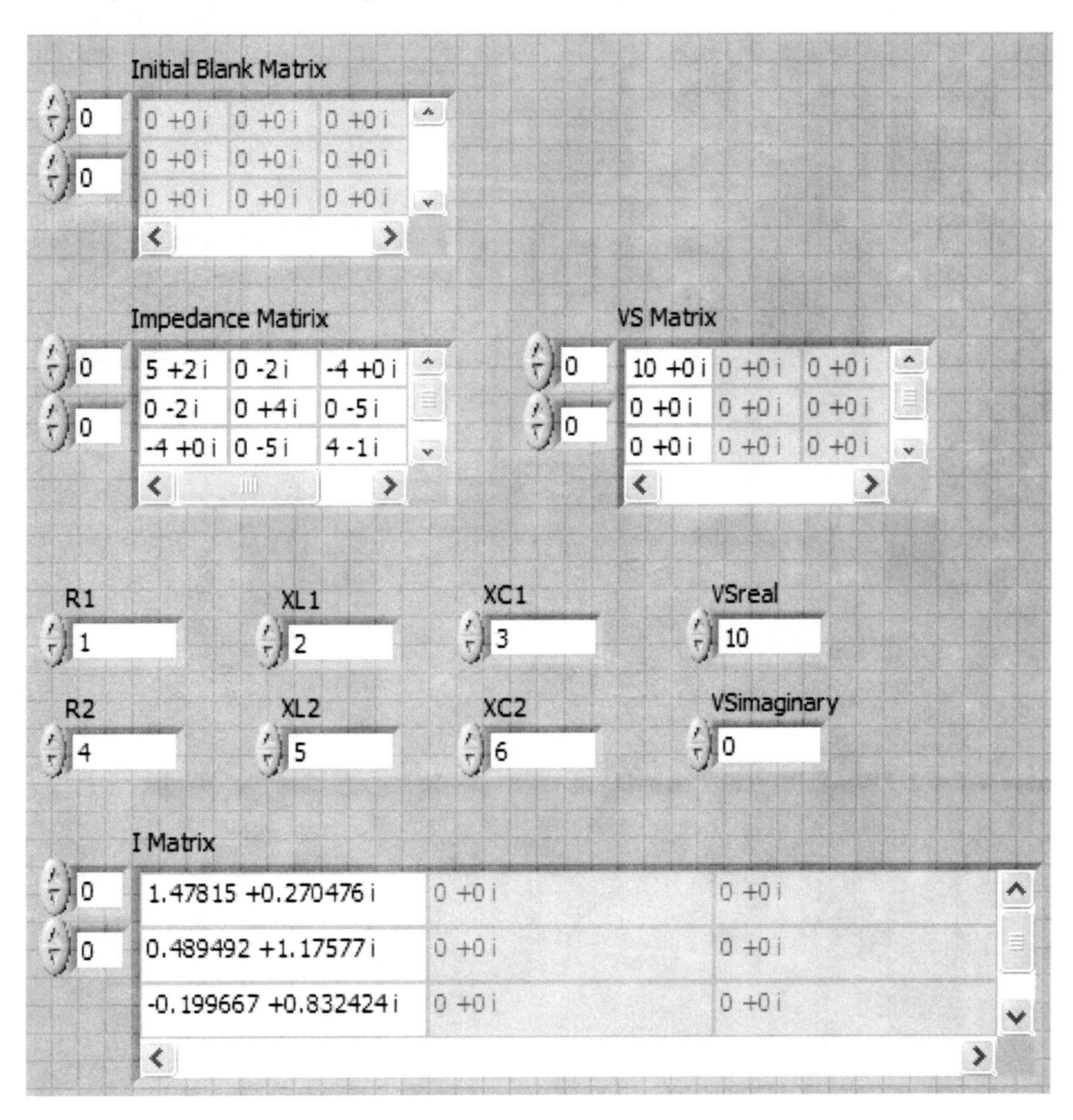

Figure 4-1-4-2 "Front Panel" with input and solution data.

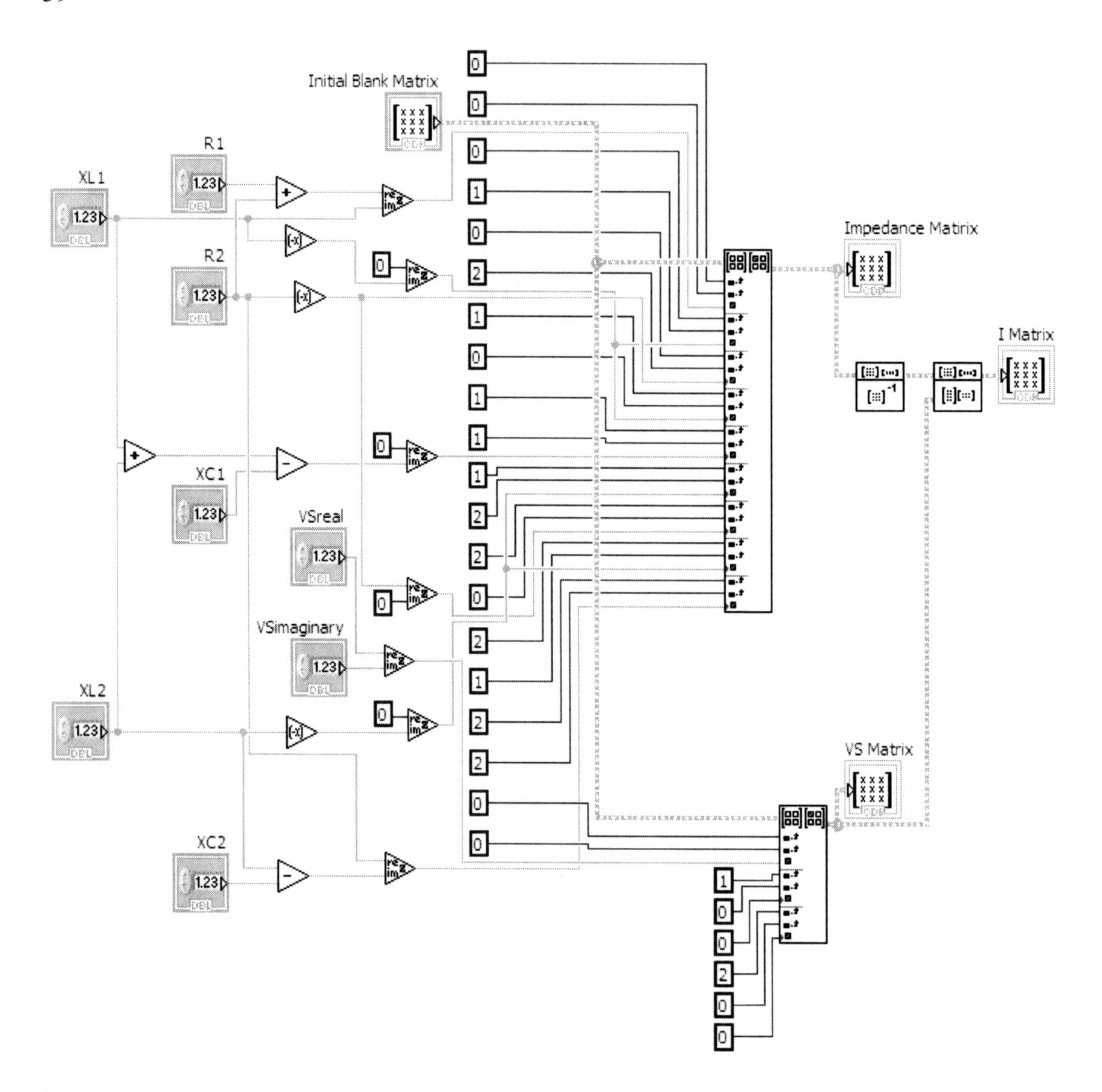

Figure 4-1-4-3 "Block Diagram" showing complex number matrix function blocks.

4.1.5 AC INDUCTION MOTOR ANALYSIS

Problem:

An AC three-phase 5 hp induction motor powers a fan. The power output of the motor is assumed to be a constant 5 hp. Using LabVIEW and the motor's equivalent circuit determine the motor's input current, power loss, and slip at input voltages of 480 and 375 Vrms.

The values for the equivalent circuit of the motor are R1 = 1.6 Ω, R2 = 1.4 Ω, RM = 980 Ω, XM = 120 Ω, X = 6.5 Ω. VS is assumed to be at angle 0. See Figure 4-1-5-1.

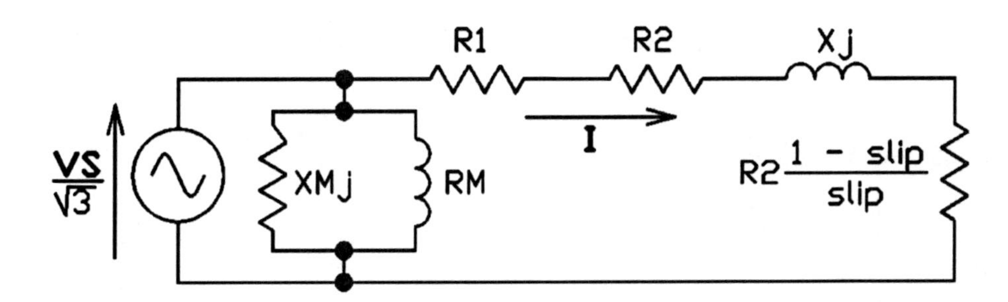

Figure 4-1-5-1 AC induction motor phasor equivalent circuit of one line-to-neutral phase.

System of equations that describes this problem:

(i) Motor output power (Hp) PO = 3·|**I**|²·R2·(1-slip)/(slip·746)

(ii) Rotor equivalent circuit current (amps) **I** = {**VS**/[SQRT(3)]}/(R1 + R2/slip + Xj)

(iii) Motor power loss (watts) PL = 3·[|**I**|²·(R1 + R2) + |**VS**/[SQRT(3)]|²/RM]

(iv) Total motor current (amps) **IT** = **I** + {**VS**/[SQRT(3)]}/RM + {**VS**/[SQRT(3)]}/(XMj)

4.1.5.1 Solution using a Purely Graphical VI

A purely graphical VI is used to solve the equations for slip. Potential slip solution values are put into the equations and the differences between the equations' outputs graphed versus slip to see how well they are solved. The difference curve crosses the zero axis at the proper slip. An advantage of this method is that if there are multiple slip answers, they will appear on the graph.

Once the graphical method has determined the slip, the programmer types the value into the VI and the other equations are solved.

LabVIEW's "For Loop" and "Waveform Graph" are demonstrated here.

Solution:

1) Equation (i) is solved for the magnitude of I.

(i') I = | {(PO·slip·746)/[3·R2·(1-slip)]}^.5 | ....This is called I1 in the VIs.

2) The equation (ii) is also solved for the magnitude of I.

(ii') I = | {VS/[SQRT(3)]}/(R1 + R2/slip + Xj) | ....This is called I2 in the VIs.

3) In the VI the I of (i') minus the I of (ii') is graphed versus slip. One hundred slip values are evaluated from .001 to .101.

4) The zero crossing slip is determined by looking at the graph. Equations (ii'), (iii), and the magnitude of equation (iv) are solved using the zero crossing slip.

5) Create the VI of Figure 4-1-5-1-2 and 4-1-5-1-3.

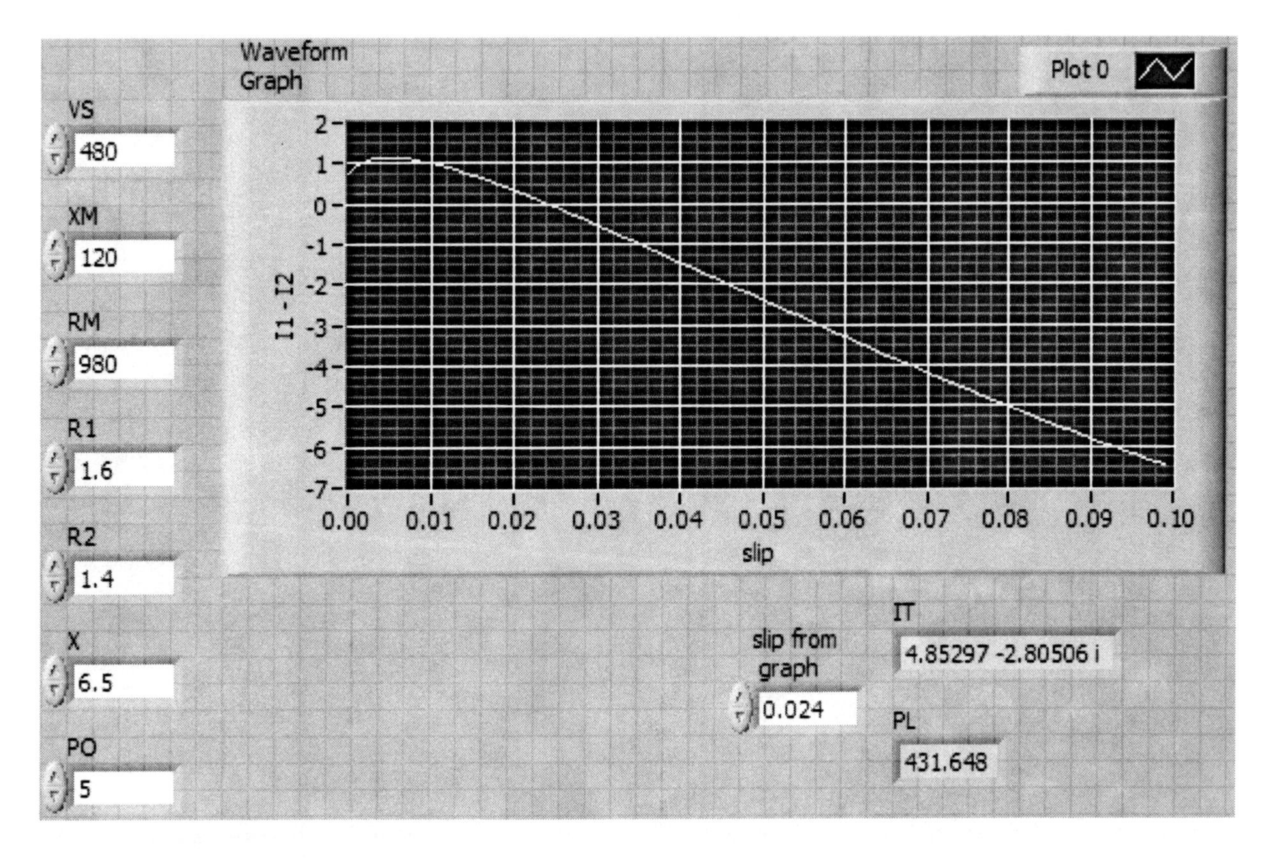

Figure 4-1-5-1-2 "Front Panel" for the purely graphical VI. The graph shows the results of the first VI run, with VS equal to 480 volts. The data under the graph is from the second VI run, after the programmer typed in the zero-crossing .024 slip value.

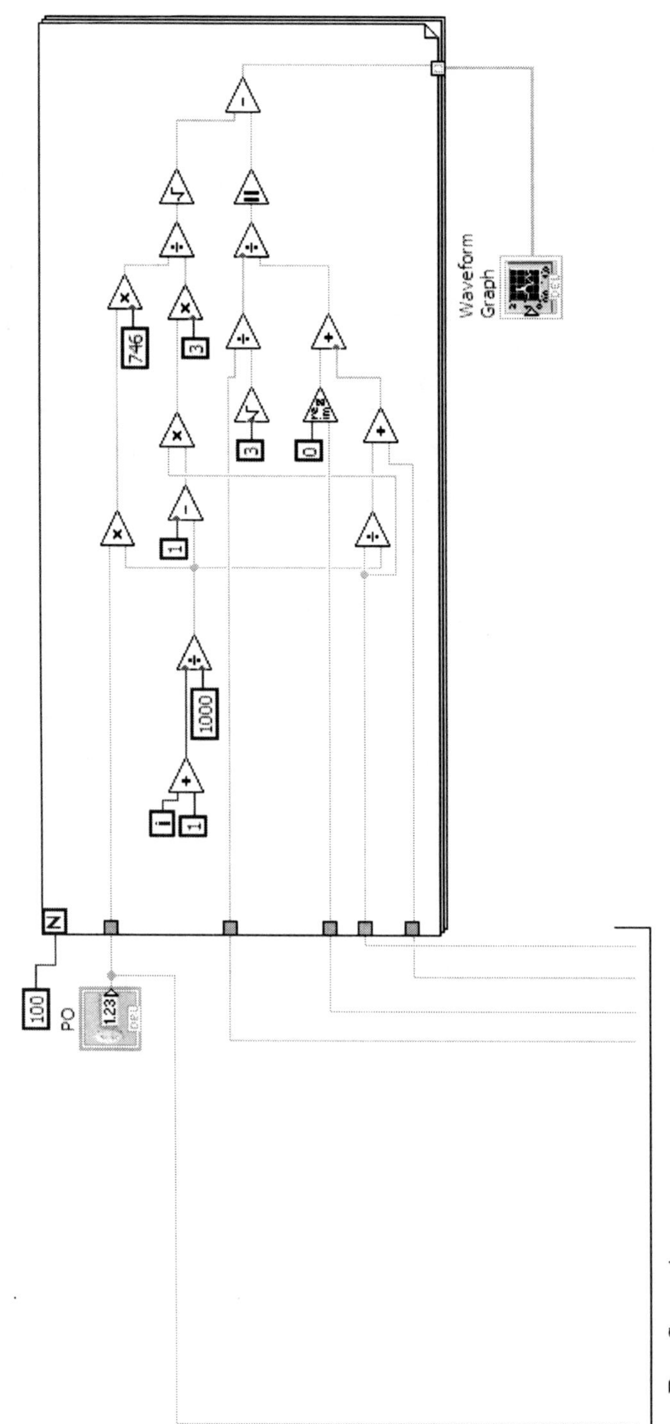

Dataflow wires connect on next page

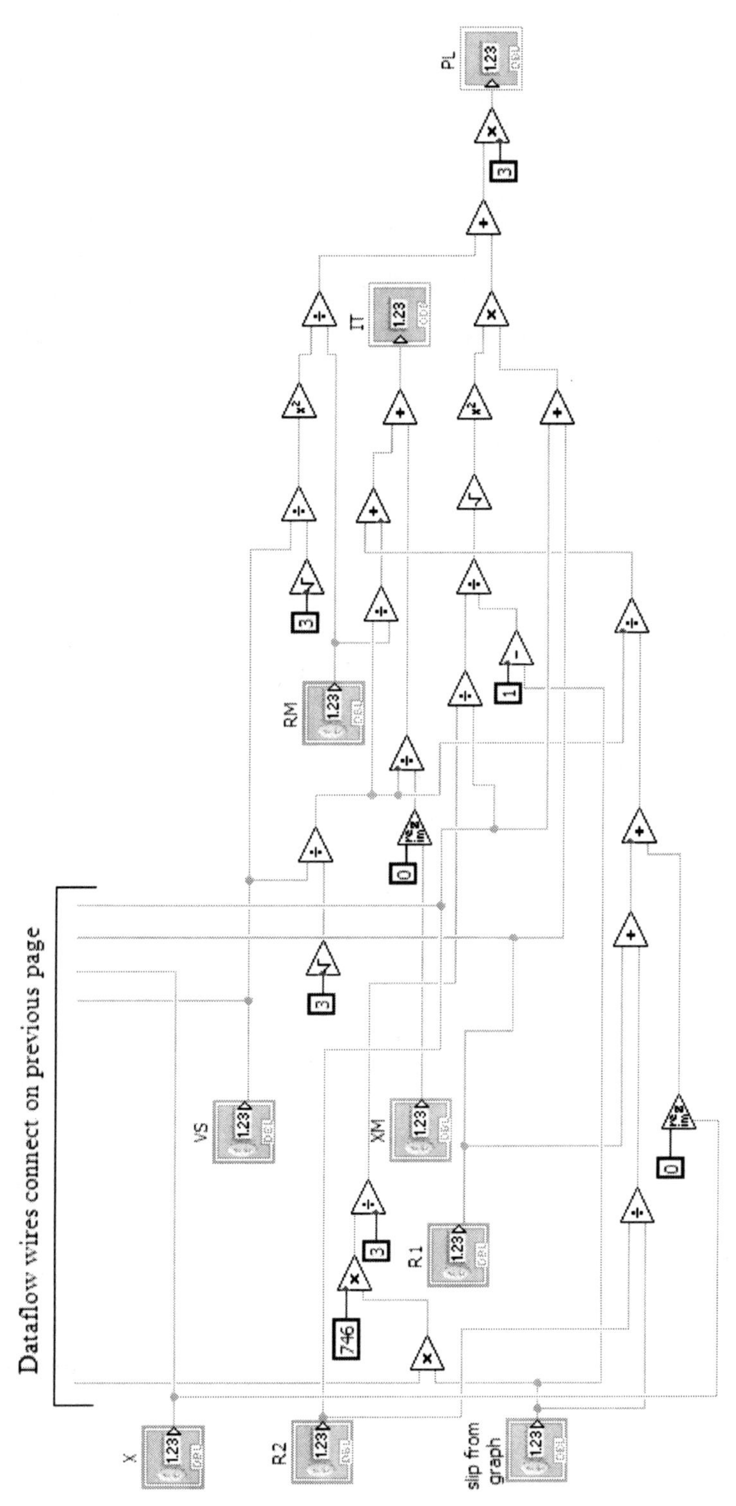

Figure 4-1-5-1-3 "Block Diagram". Page 43 shows the "For Loop", PO input block, "Waveform Graph" output block, and function blocks. Page 44 shows input blocks, PL and IT output blocks, and function blocks.

6) Put the "For Loop" on the "Block Diagram" by left-clicking on "View", "Functions Palette", "Programming", and "Structures". Left-click and drag the "For Loop" onto the "Block Diagram". Set the value of "N" to 100 by left-clicking on "View", "Tools Palette", and "Operate Value". Then place the cursor over the "N", right-click, left-click on "Create Constant", and type in 100.

7) The slip is represented in the "For Loop" of the VI as the iteration "i" divided by 1000. The iterations go from 0 to 100. A one (1) is added to the iterations so that the first iteration, "i" = 0, does not cause a divide by zero error in the VI. When the division takes place, the VI sees the slip varying from 1/1000 to 101/1000. At each iteration, the "For Loop" will do the calculations contained inside of it.

8) On the "Front Panel" get the "Waveform Graph" output block by left-clicking on "View", "Controls Palette", "Modern", "Graph". Then left-click and drag the "Waveform Graph" output block onto "Front Panel".

9) On the "Front Panel", change the "Waveform Graph" output block axis labels by left-clicking on "View", "Tools Palette", and "Edit Text". Then left-click when the cursor is on each axis label, delete and backspace erase the existing label and type in the new labels, "I1 – I2" for vertical and "slip" for horizontal.

10) By default the "Waveform Graph" output block chooses values that will contain "slip" from 0 to 100 and the "I1 – I2" from -10 to + 10.

11) The "Waveform Graph" output block "slip" axis should be scaled so that it displays slip values rather than slip x 1000 values. To scale the "slip" axis right-click anywhere on the "Waveform Graph" output block. Next, left-click on "Properties", and "Scales". Type in .001 for the "Scaling Factors" "Multiplier". It will also be necessary to change the number of digits displayed after the decimal point in the "slip" axis figures. To do this left-click on the "Display Format" tab and select 2 "digits". Finally, left-click on "OK".

12) The "Waveform Graph" output block "I1 – I2" and "slip" axis ranges are overwritten by left-clicking on "View", "Tools Palette", and "Edit Text". Then, when the cursor is on each axis limit, left-click, delete, and backspace to erase the existing number and type in the desired.

13) Values for the magnitudes of PL and IT are determined for the slip value that was found with the "For Loop". The VI solves equation (iii) and (iv) for magnitudes. The equations graphical equivalents appear in the VI in Figure 4-1-5-1-3 on page 44.

14) The size and complexity of this example's graphical VI demonstrates that LabVIEW can be confusing with larger graphical VIs. On VIs this complex, it would be best to use LabVIEW with its MathScript toolbox. With MathScript, LabVIEW handles equations as text in lines, as is done in programs like Mathcad, MATLAB, FORTRAN, etc.

15) The VI is run again with VS equal to 375 volts. See Figure 4-1-5-1-4.

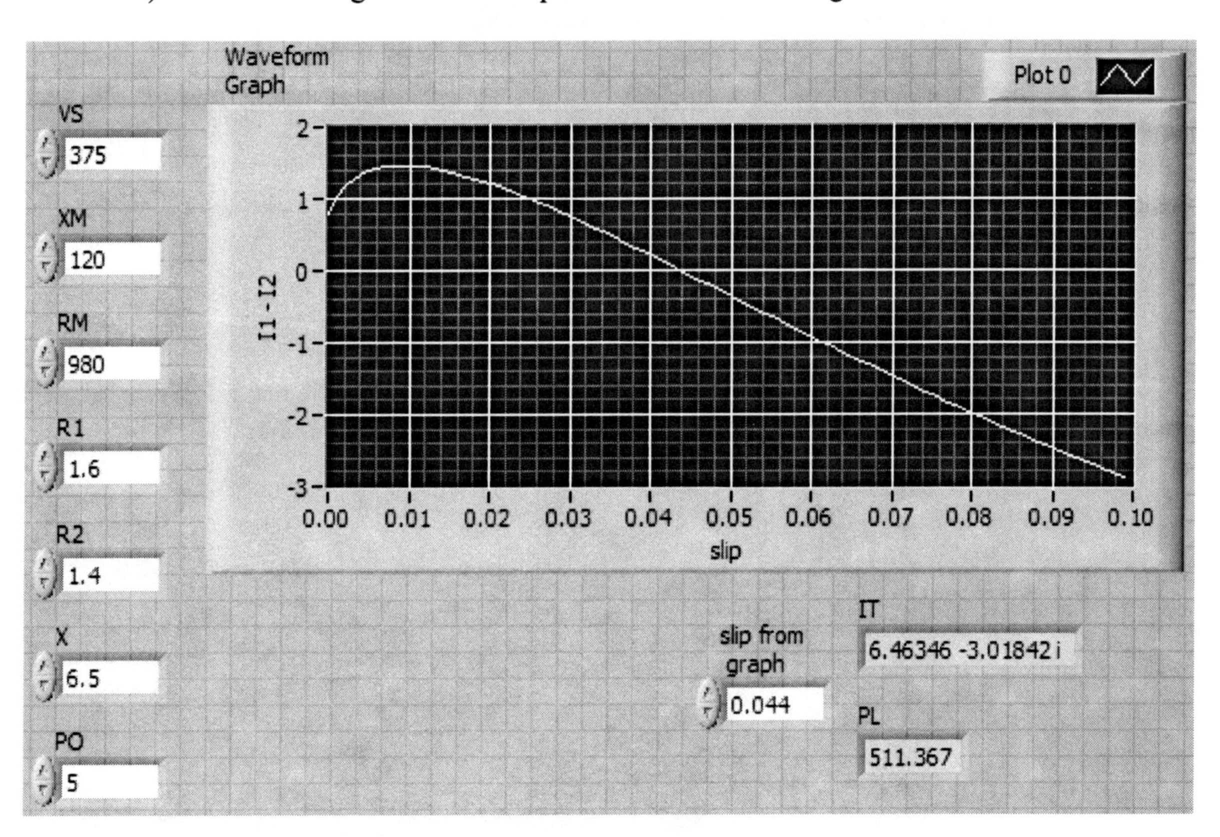

Figure 4-1-5-1-4 "Front Panel". The graph shows the results of the first VI, run with VS equal to 375 volts. The data under the graph is from the second VI run, after the programmer typed in the .044 zero-crossing slip value.

4.1.5.2 Solution using "Formula Node" function blocks in a VI

An alternate method of solution is to use "Formula Node" function blocks to evaluate the equations, rather than graphical programming. The equations are written into the "Formula Node" function blocks.

Three "Formula Node" function blocks will be created.

Note that "Formula Node" function blocks can only evaluate equations formed with real numbers.

Solution:

1) Create a new VI.

2) The same "For Loop" is used as before.

3) Block and "Ctrl c" copy the "Front Panel" material of the VI of Figure 4-1-5-1-2. "Ctrl v" place the material on the "Front Panel" of Figure 4-1-5-2-1. See Figure 4-1-5-2-1.

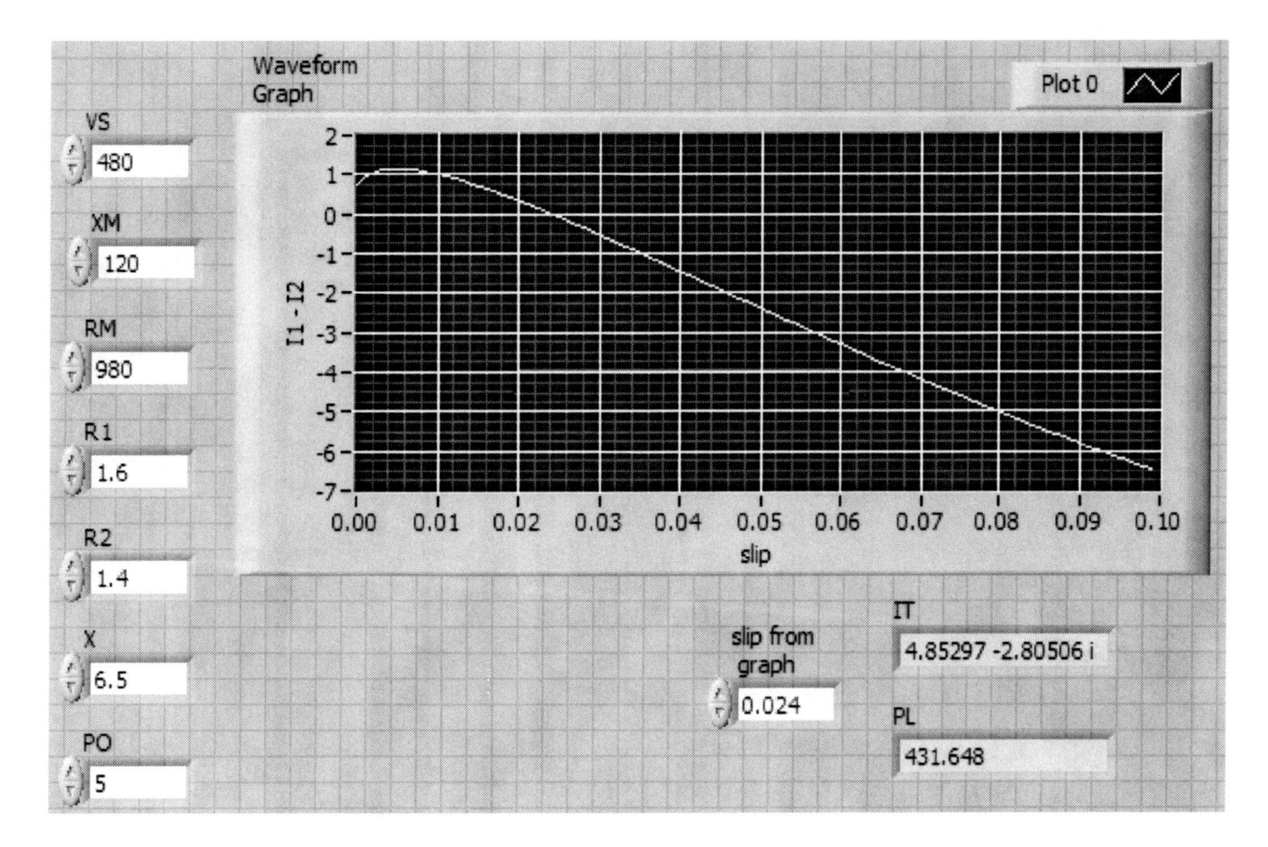

Figure 4-1-5-2-1 "Front Panel" of the "Formula Node" VI. The graph shows the results of the first VI run, with VS equal to 480 volts. The data under the graph is from the second VI run, after the programmer typed in the .024 slip value.

4) Inside the "For Loop" create a "Formula Node" function block. To do this left-click on "View", "Functions Palette", "Programming", and "Structures Palette". Then left-click and drag the "Formula Node" function block into the "For Loop" box. See the "Block Diagram" in Figure 4-1-5-2-2.

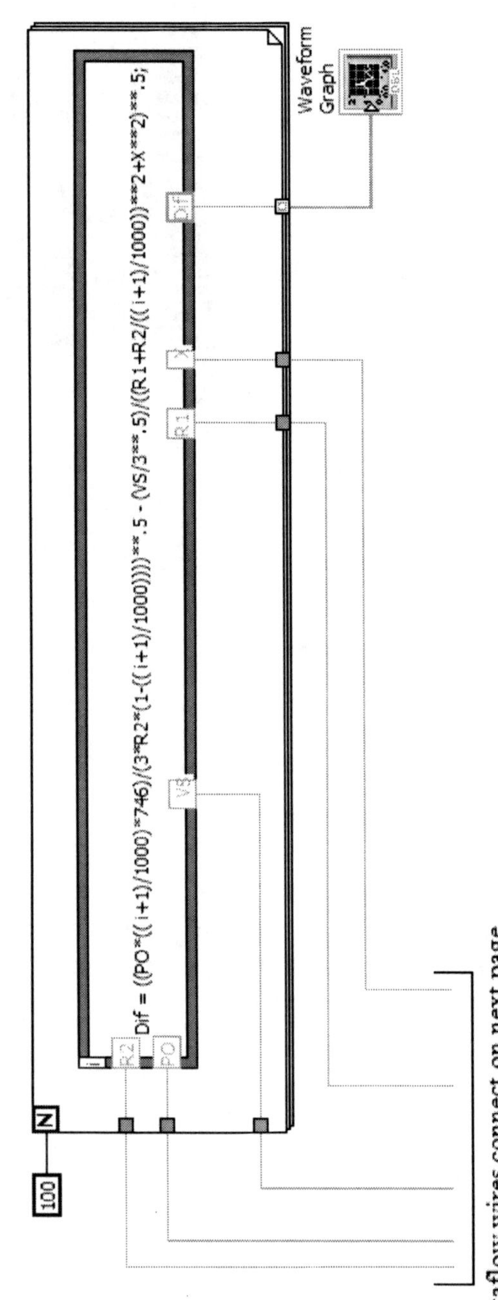

Dataflow wires connect on next page

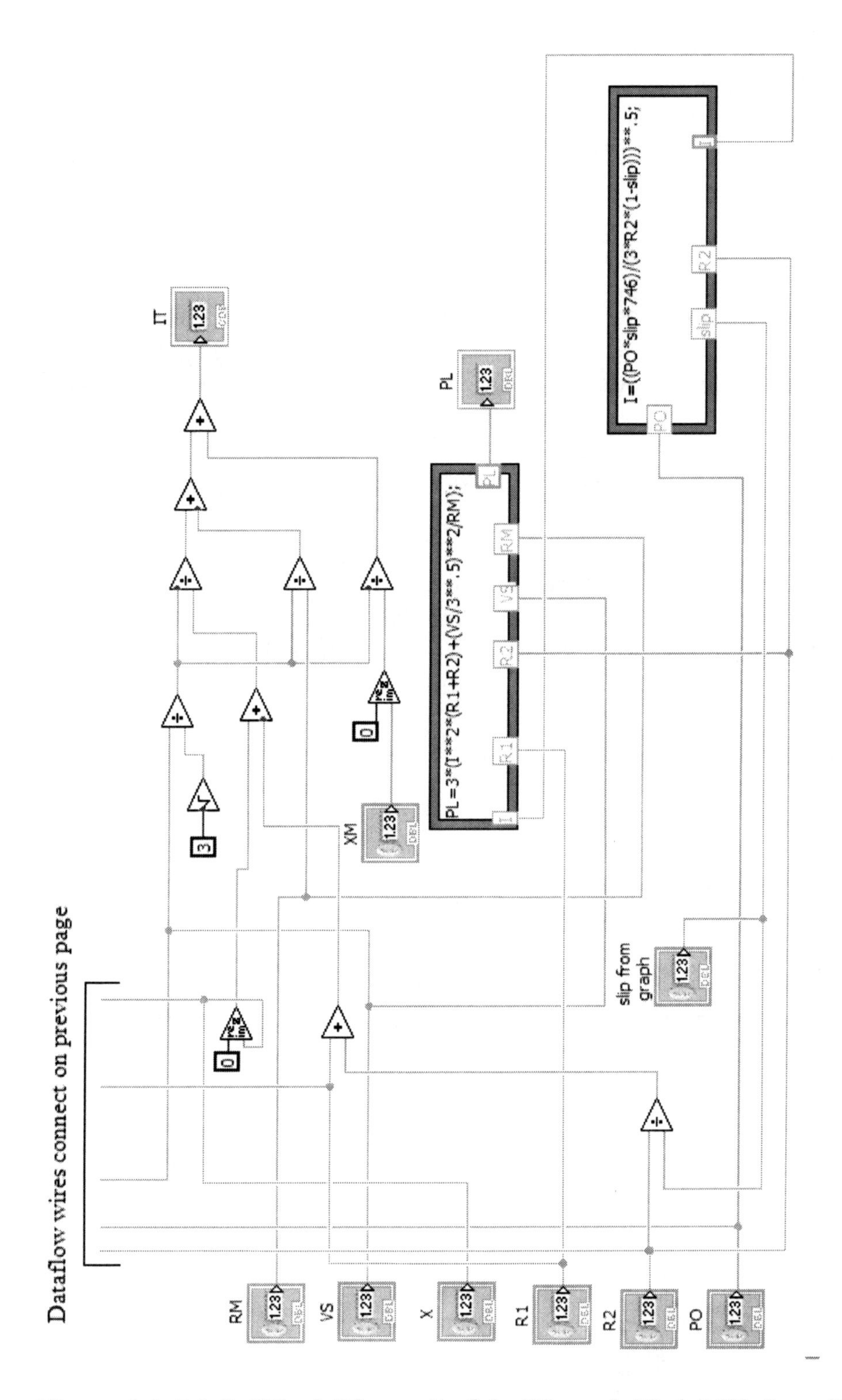

Figure 4-1-5-2-2 "Block Diagram" of the "Formula Node" VI. Page 49 shows the "For Loop" and "Waveform Graph" output block. Page 50 shows the input data and solution data.

5) Using the "Tool Palette" "Position/Size/Select" tool drag the "For Loop" and "Formula Node" function block out to make them larger.

6) From page 41, the equation for the difference between I1 from equation (i') and I2 from equation (ii') is:

I1 – I2 = | {(PO·slip·746)/[3·R2·(1-slip)]}^.5 | - | {VS/[SQRT(3)]}/(R1 + R2/slip + Xj) |

7) "Formula Node" function blocks cannot evaluate imaginary numbers. Therefore, the equation is changed to a form with only real numbers. Also, "I1 – I2" is changed to "Dif", so that its "Formula Node" function block produces a single variable. Written in "Formula Node" format the equation is:

```
Dif = ((PO*((i+1)/1000)*746)/(3*R2*(1-((i+1)/1000))))**.5 –
        (VS/3**.5)/((R1+R2/((i+1)/1000))**2+X**2)**.5;
```

Type this equation into the "Formula Node" by left-clicking on "Operate Value" in the "Tools Palette" and left-clicking in the "Formula Node". Note that the "Formula Node" function block requires that there is just one output variable on the left side of the equals sign and that the equation is followed by a semi-colon.

8) To create input terminals for each equation variable right-click on the "Formula Node" border and left-click on "Add Input". Label the input terminal with variable labels, i.e. R1, R2, VS, etc.

9) Connect the input terminals to their appropriate input blocks. The input blocks are already on the "Block Diagram", since they were automatically placed there when the "Front Panel" was created.

10) Create an output terminal by right-clicking on the "Formula Node" function block border and left-clicking on "Add Output". Label the output terminal Dif and connect it to the "Waveform Graph" output block.

11) The equations for PL and IT can also be solved using a "Formula Node" function block. Again, "Formula Node" function block equations cannot include complex numbers. The equation for motor power loss, equation (iii'), is relatively easy to write.

(iii') $PL = 3\cdot[I^2\cdot(R1 + R2) + (VS/[SQRT(3)])^2/RM]$

In the "Formula Node" format it is:

```
PL = 3*(I**2*(R1 + R2) + (VS/3**.5)**2/RM);
```

Here I is found with equation (i'):

$I = \left| \{(PO\cdot slip\cdot 746)/[3\cdot R2\cdot(1\text{-}slip)]\}\text{^}.5 \right|$

In the "Formula Node" format it is:

```
I = ((PO*slip*746)/(3*R2*(1-slip)))**.5;
```

4.1.6 3D GRAPH OF RLC FILTER VOLTAGE OUTPUT/INPUT VERSUS FREQUENCY AND LOAD RESISTANCE

LabVIEW can make 3D Scatter, Bar, Pie, Stem, Ribbon, Contour, Quiver, Comet, Surface, Mesh, and Waterfall graphs. It can also make images of moving 3D objects and scenes.

Problem:

Use LabVIEW to create a "3D Parametric Surface" graph of AC filter voltage output/input versus frequency and load resistance. The filter's capacitance is 10 μF, inductance is 100 μH, and effective series resistance is 8 Ω. The frequency is varied from 2 to 14,252 Hz. The load resistance is varied from .1 to 19.1 Ω. The circuit is shown in Figure 4-1-6-1.

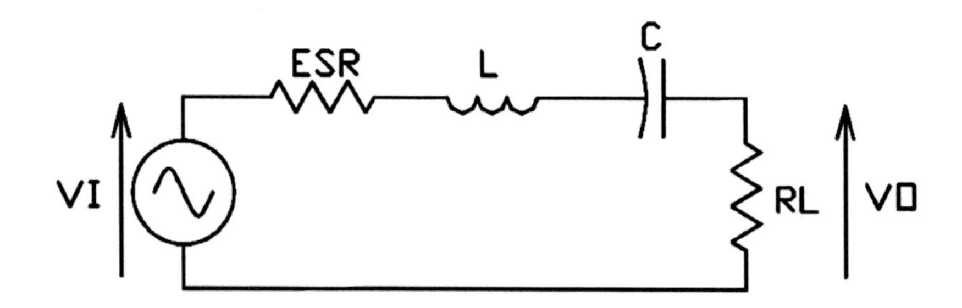

Figure 4-1-6-1 Filter circuit.

LabVIEW's nested "For Loop"s and a 3D parametric surface graph are demonstrated here.

Solution:

1) The equation is:

$$|VO/VI| = RL/[(RL + ESR)^2 + \{2\cdot\pi\cdot F\cdot L - 1/(2\cdot\pi\cdot F\cdot C)\}^2]^{0.5}$$

2) To make the VI easier to write and read the equation will be changed to:

$$|VO/VI| = RL/(A^2 + B^2)^{0.5}$$

Where:

$$A = RL + ESR$$

$$B = 2\cdot\pi\cdot F\cdot L - 1/(2\cdot\pi\cdot F\cdot C)$$

3) Write the VI shown in Figure 4-1-6-2.

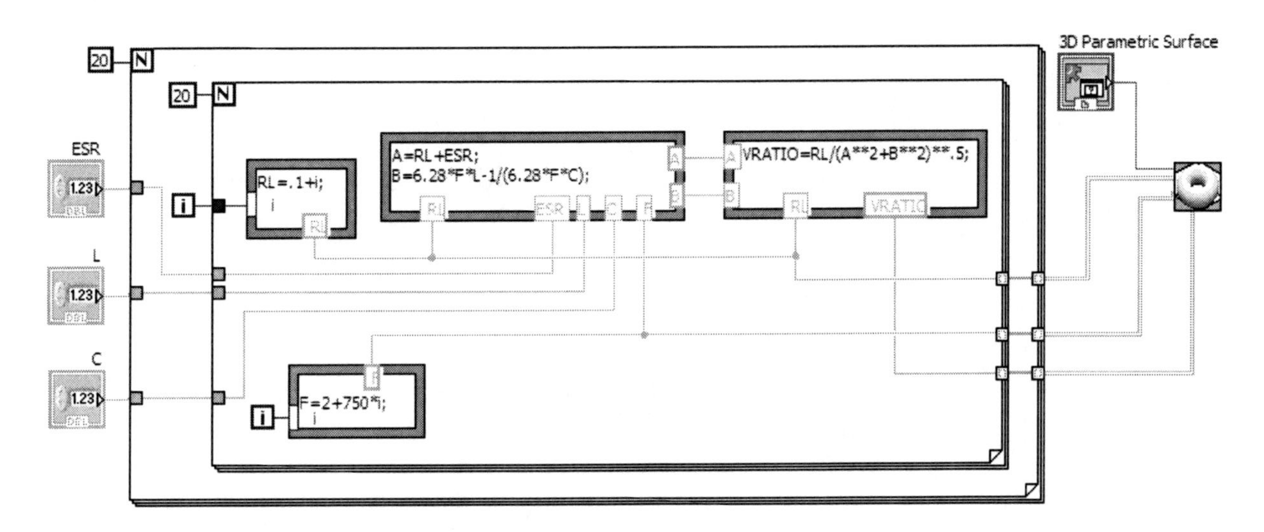

Figure 4-1-6-2 "Block Diagram" showing nested "For Loop"s and "3D Parametric Surface" output blocks.

4) The "For Loop"s are nested. At first, the outer "For Loop" is at its minimum value of 0. This value is used in the resistance equation, "RL=.1+i;". The RL value is used in the circuit equations. The inner "For Loop" goes through 20 steps from 0 to 19. Each step value is used in the frequency calculating equation, "F=2+750*i;". The frequencies are used in the circuit equations.

After the inner "For Loop" has completed its 20 steps the outer "For Loop" advances one step so that the calculations are repeated with newer RL values.

5) LabVIEW automatically stores the values it calculated for RL, F, and |VO/VI|. and makes them available to the "ActiveV 3D Parametric Graph.vi" output block. Create the "ActiveV 3D Parametric Graph.vi" in the "Front Panel" by left-clicking on "View", "Controls Palette", "Classic", and "Classic Graph". Left-click, hold and drag "ActiveV 3D Parametric Graph.vi" onto the "Front Panel". Left-click, drag out and stretch out the graph to a larger size. See the "Front Panel" in Figure 4-1-6-3.

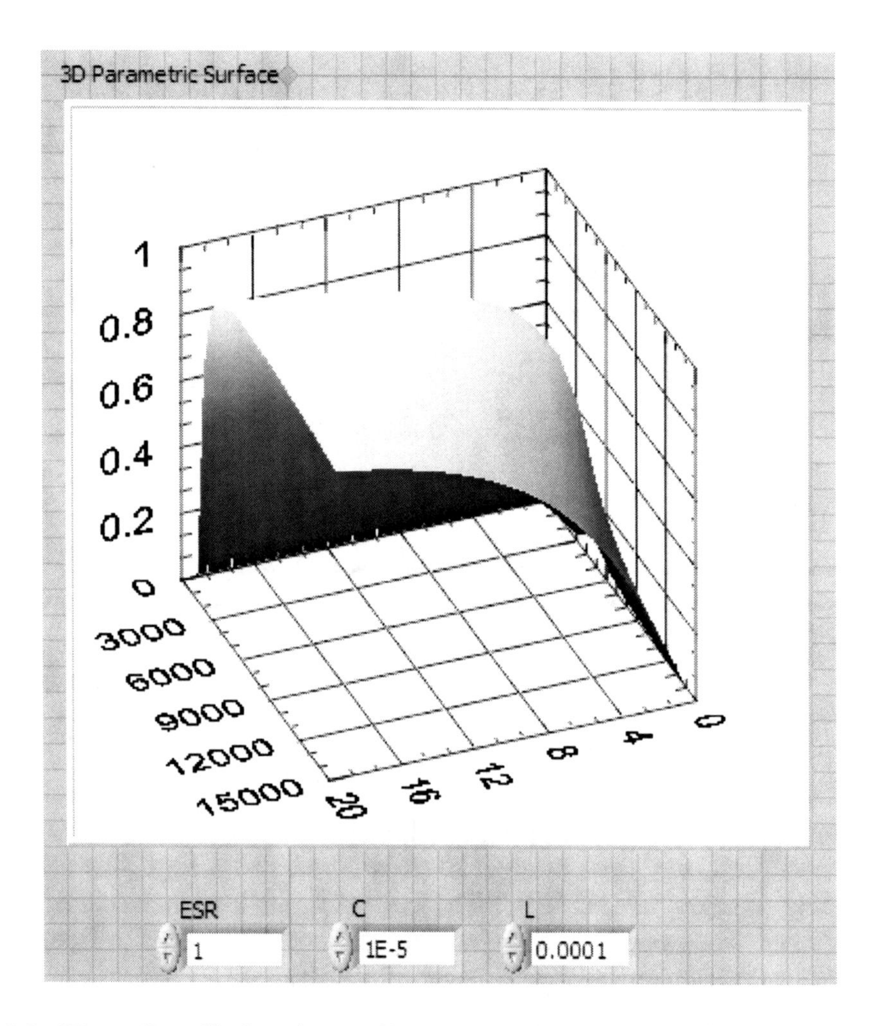

Figure 4-1-6-3 "Front Panel" showing a "3D Parametric Surface" plot.

6) In the "Block Diagram", connect the RL, F, and |VO/VI|. dataflow wires from "Formula Node" function blocks to the "3D Parametric Surface VI" function block X, Y, and Z inputs.

7) Once the VI is run the 3D graph can be rotated to different views. To do this, go to the "Front Panel" and left-click on the "Tools Palette" and "Operate Value". Then place the cursor on the 3D graph and press and hold down the left mouse button. Move the cursor to rotate the graph to the desired view. During rotation, the individual calculated points are displayed rather than a smoothed surface.

8) Axis labels can be added. Put the cursor on the 3D graph and right-click. Then left-click on "CW Graph 3D Control", "Properties", and "Axes". Select each axis and type in the desired labels. The rotated and axes-added "Front Panel" can be seen in Figure 4-1-6-4.

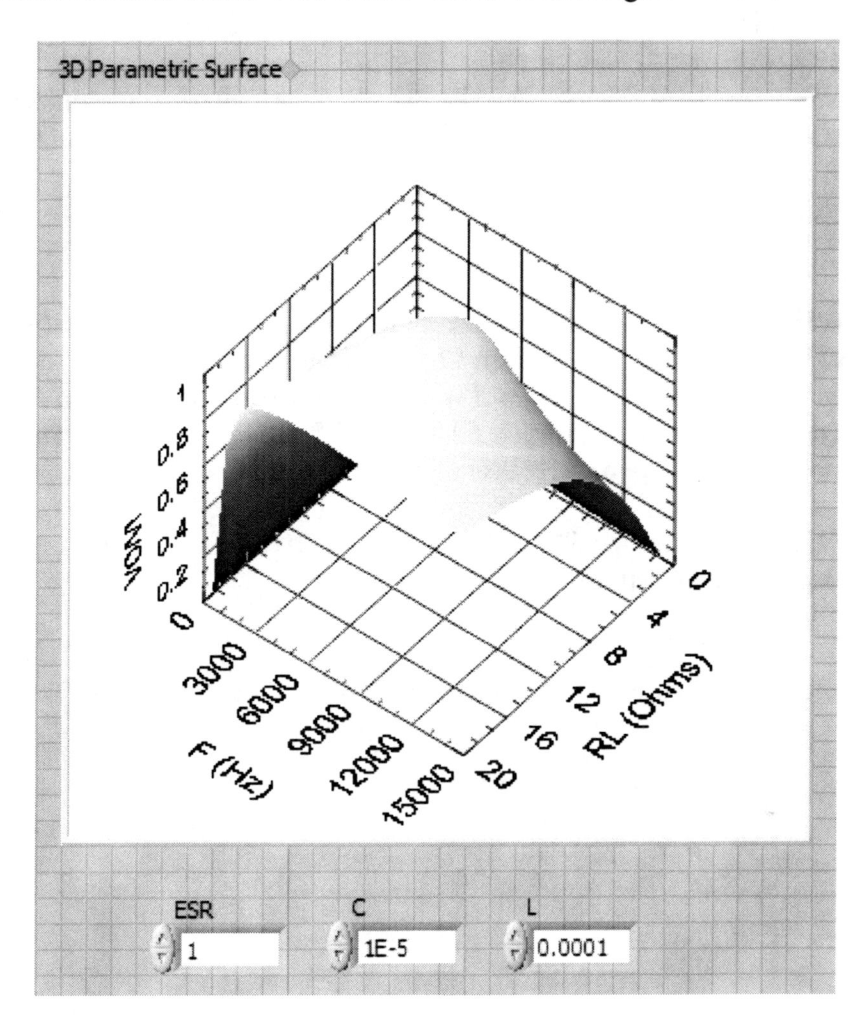

Figure 4-1-6-4 "Front Panel" after rotation and addition of axis labels.

4.2 REAL-TIME VIs WITH SIMULATED SIGNALS

The "Simulated Signal" function block can create signals internally so that LabVIEW can do the same real-time analyses as can be done with a DAQ and physical input signals.

LabVIEW's "Simulated Signal" function block is useful to a person learning to use LabVIEW and to an experienced LabVIEW programmer who wants to test a signal analysis VI.

4.2.1 FILTERING OF A SIMULATED SIGNAL

Problem:

Use LabVIEW to filter a LabVIEW simulated signal. Display the unfiltered simulated signal and filtered signal in "Front Panel" "Waveform Chart" output blocks. The signal should be a 60 Hz sine wave with an amplitude of 1 and a rider of white noise with an amplitude of .2. Filtered outputs should be displayed for frequencies between 59.5 and 60.5 Hz in one "Waveform Chart" output block and for all frequencies above 62.5 Hz in another "Waveform Chart" output block.

LabVIEW's "Simulate Signal" function block, "Filter" function block, and "Waveform Chart" output block are demonstrated here.

Solution:

1) Create the VI "Front Panel" shown in Figure 4-2-1-1 and the "Block Diagram" shown in Figure 4-2-1-2.

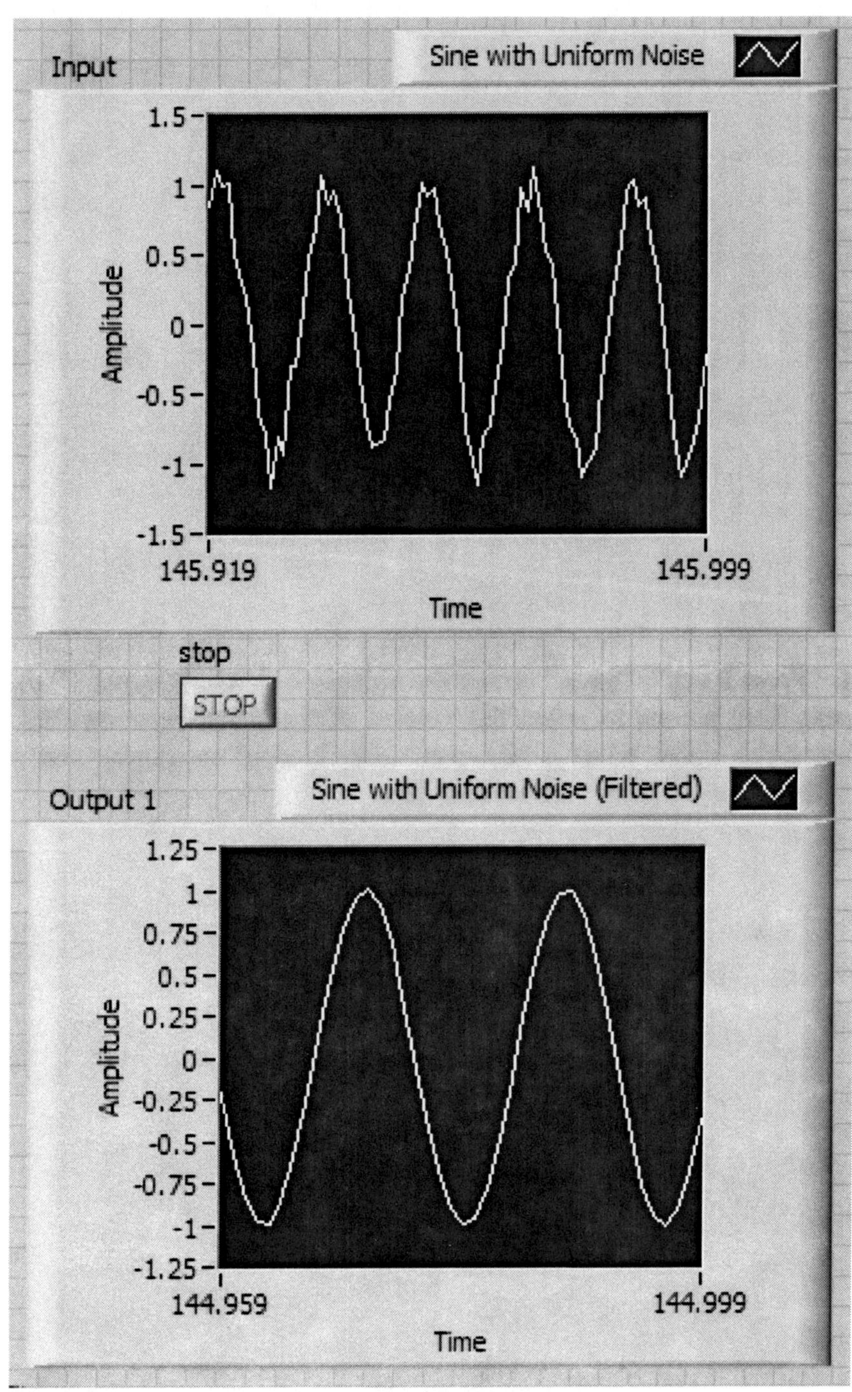
Input
Sine with Uniform Noise
1.5
1
0.5
0
-0.5
-1
-1.5
Amplitude
145.919
145.999
Time
stop
STOP
Output 1
Sine with Uniform Noise (Filtered)
1.25
1
0.75
0.5
0.25
0
-0.25
-0.5
-0.75
-1
-1.25
Amplitude
144.959
144.999
Time

Continued

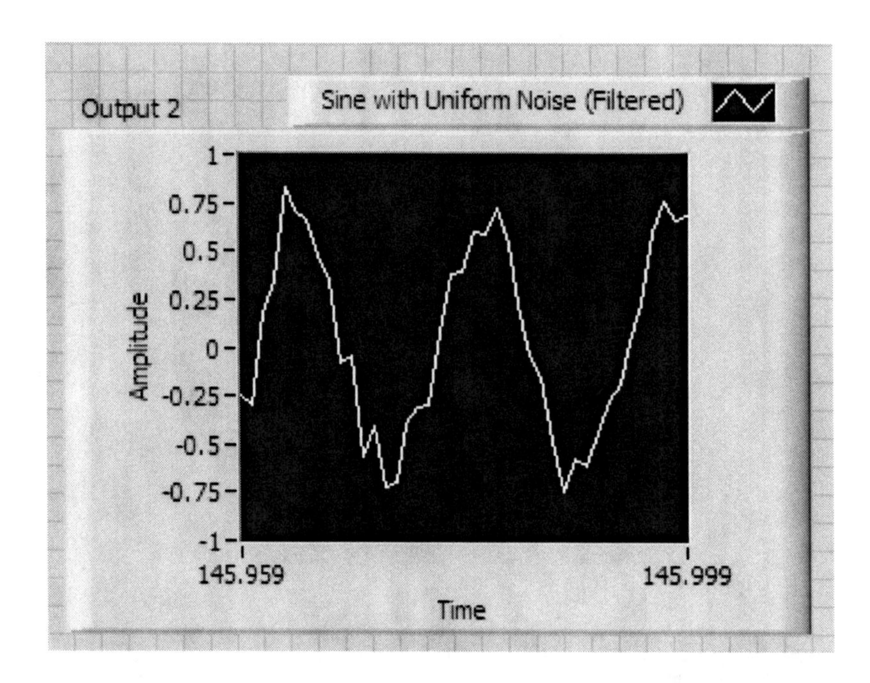

Figure 4-2-1-1 "Front Panel". "Input" shows the unfiltered signal. "Output 1" shows the signal when a band-pass filter is used to select the portion of the signal between 59.5 and 60.5 Hz. "Output 2" shows the signal when a high-pass filter is used to display the portion of the simulated signal that has a frequency above 62.5 Hz.

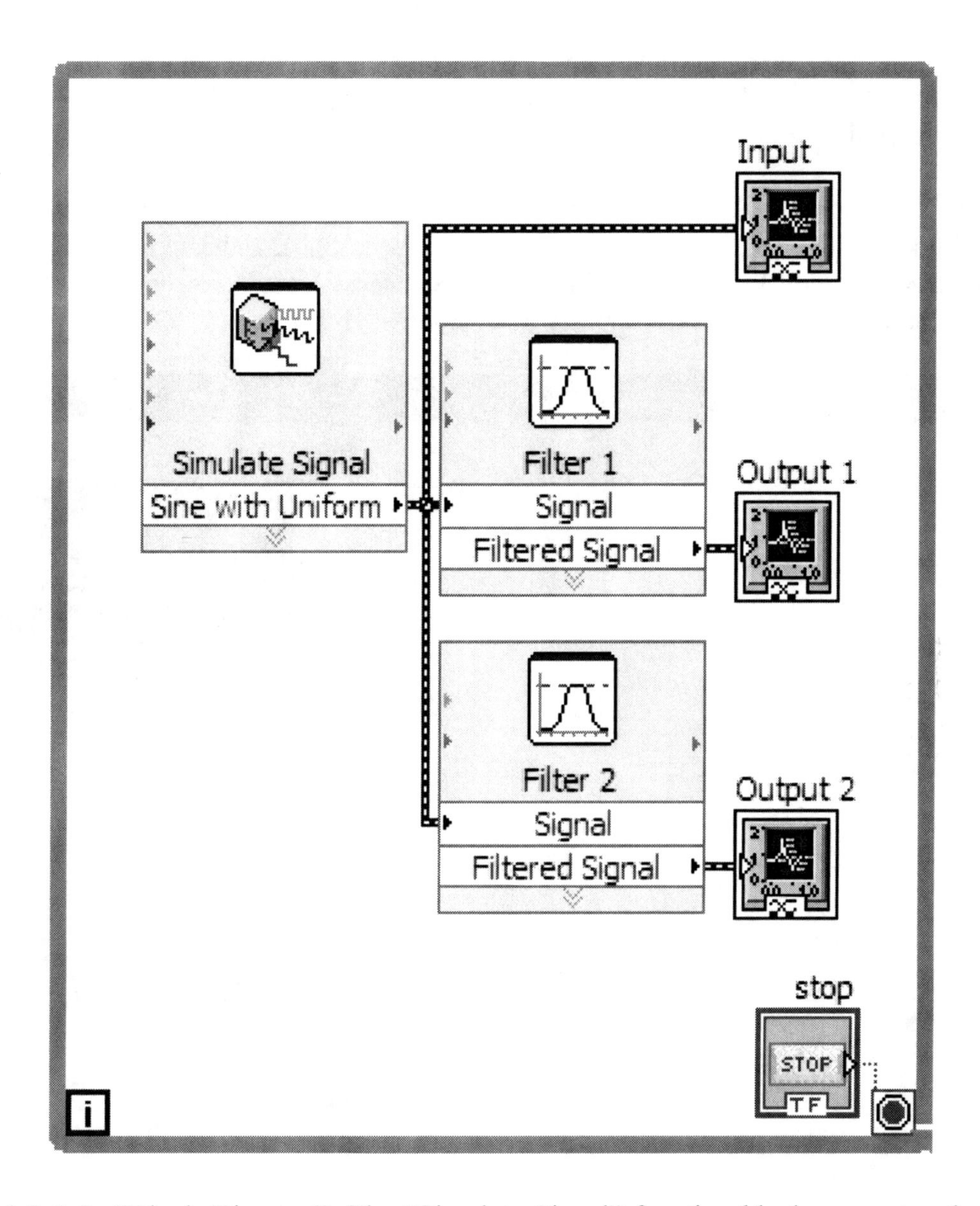

Figure 4-2-1-2 "Block Diagram". The "Simulate Signal" function block generates the signal. "Filter 1" function block is a band-pass filter that passes the portion of the signal between 59.5 and 60.5 Hz. "Filter 2" function block is a high-pass filter that passes the portion of the signal above 62.5 Hz.

2) This VI operates continuously inside a "While Loop" until one of the "Stop" buttons is left-clicked.

3) Create the "While Loop" by left-clicking on "View", "Functions Palette", "Express", and "Execution Control". Left-click, hold and drag the "While Loop" onto the "Front Panel". Drag and stretch the "While Loop" out.

4) Create the "Simulated Signal" function block in the "Block Diagram" by left-clicking on "View", "Functions Palette", "Express", and "Input". Left-click, hold and drag the "Simulate Signal" function block into the "While Loop". When the "Simulated Signal" function block is placed a "Configure Simulate Signal" window appears. See Figure 4-2-1-3.

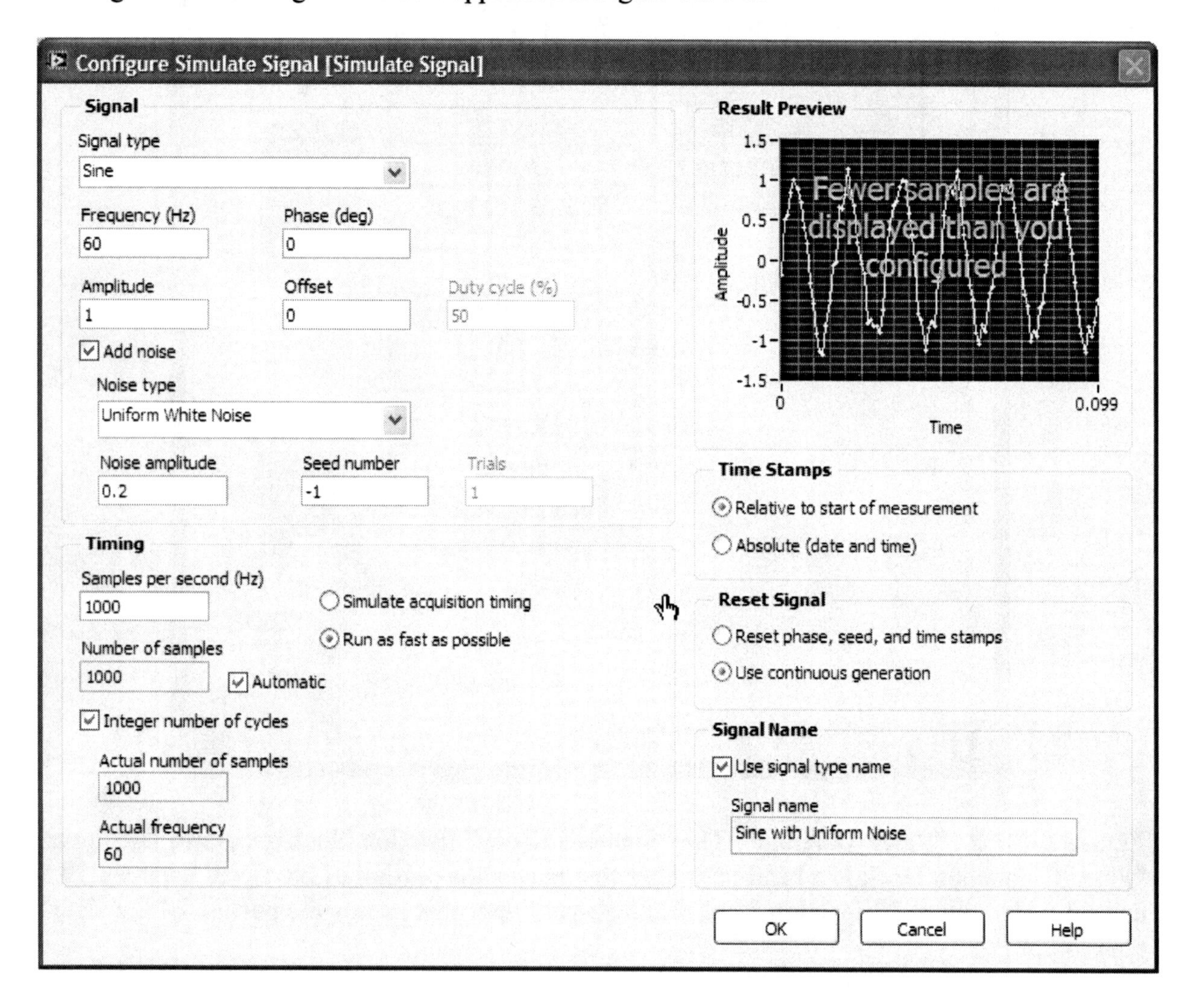

Figure 4-2-1-3 Parameters for the "Input" simulated signal.

5) In the "Configure Simulate Signal" set the "Frequency (Hz)" to 60, left-click "Add Noise", set the "Noise Amplitude" to .2, left-click on "Integer number of cycles", and set the "Number of samples" to 1000. If desired, the "Configure Simulate Signal" window can be re-opened later by right-clicking on the "Simulate Signal" function block and left-clicking on "Properties".

6) “Filter” function blocks are placed in the “Block Diagram”. The “Filter” function blocks are obtained by left-clicking on “View”, “Functions Palette”, “Express”, and “Signal Analysis”. Left-click, hold and drag the “Filter” function blocks into the “While Loop”. Each “Filter” parameter setting window is selected by right-clicking on the “Filter” function block and left-clicking on “Properties”. See Figures 4-2-1-4 and 4-2-1-5.

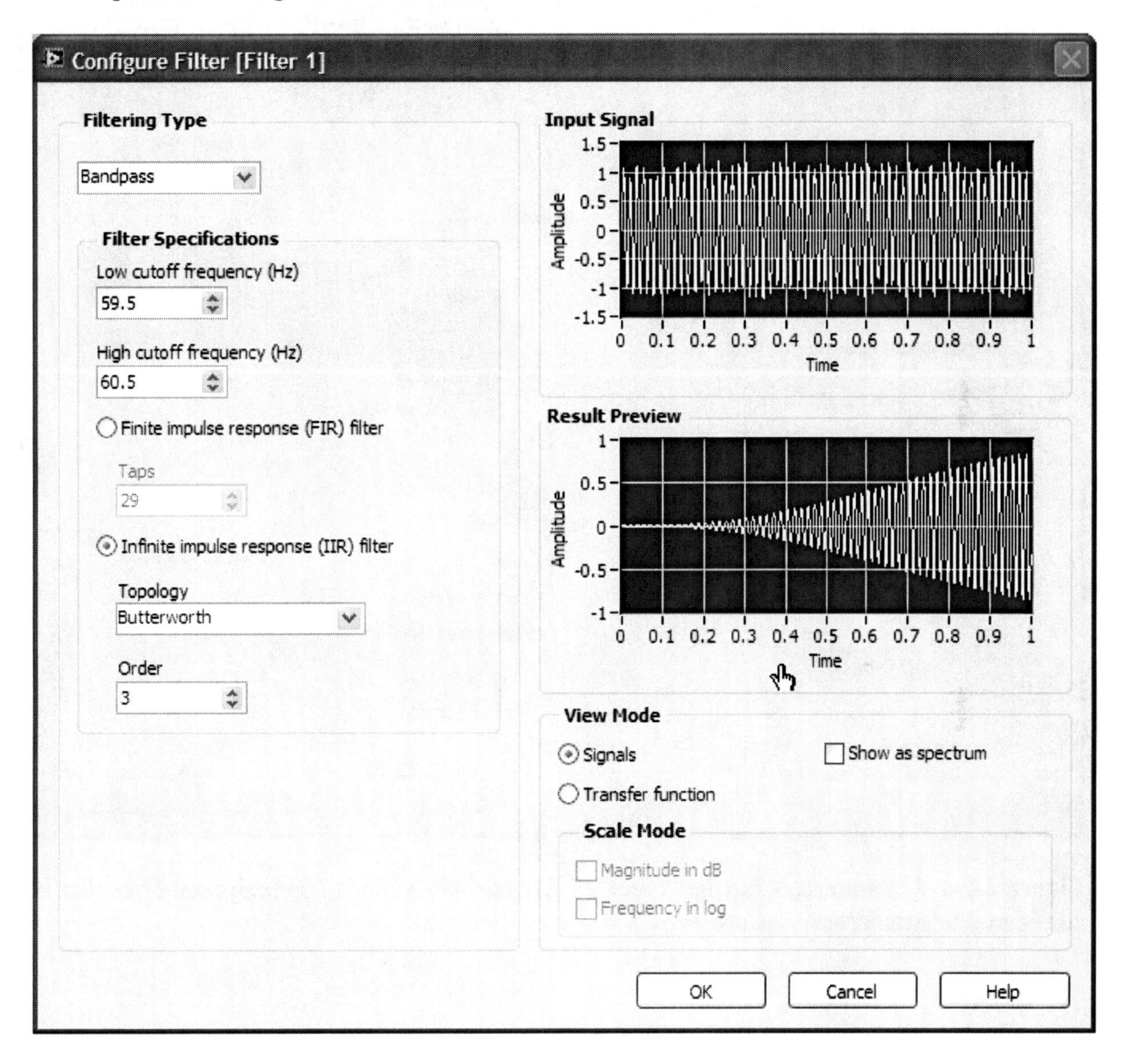

Figure 4-2-1-4 Parameters for the “Filter 1” function block. This is a band-pass filter that is configured to pass frequencies between 59.5 and 60.5 Hz.

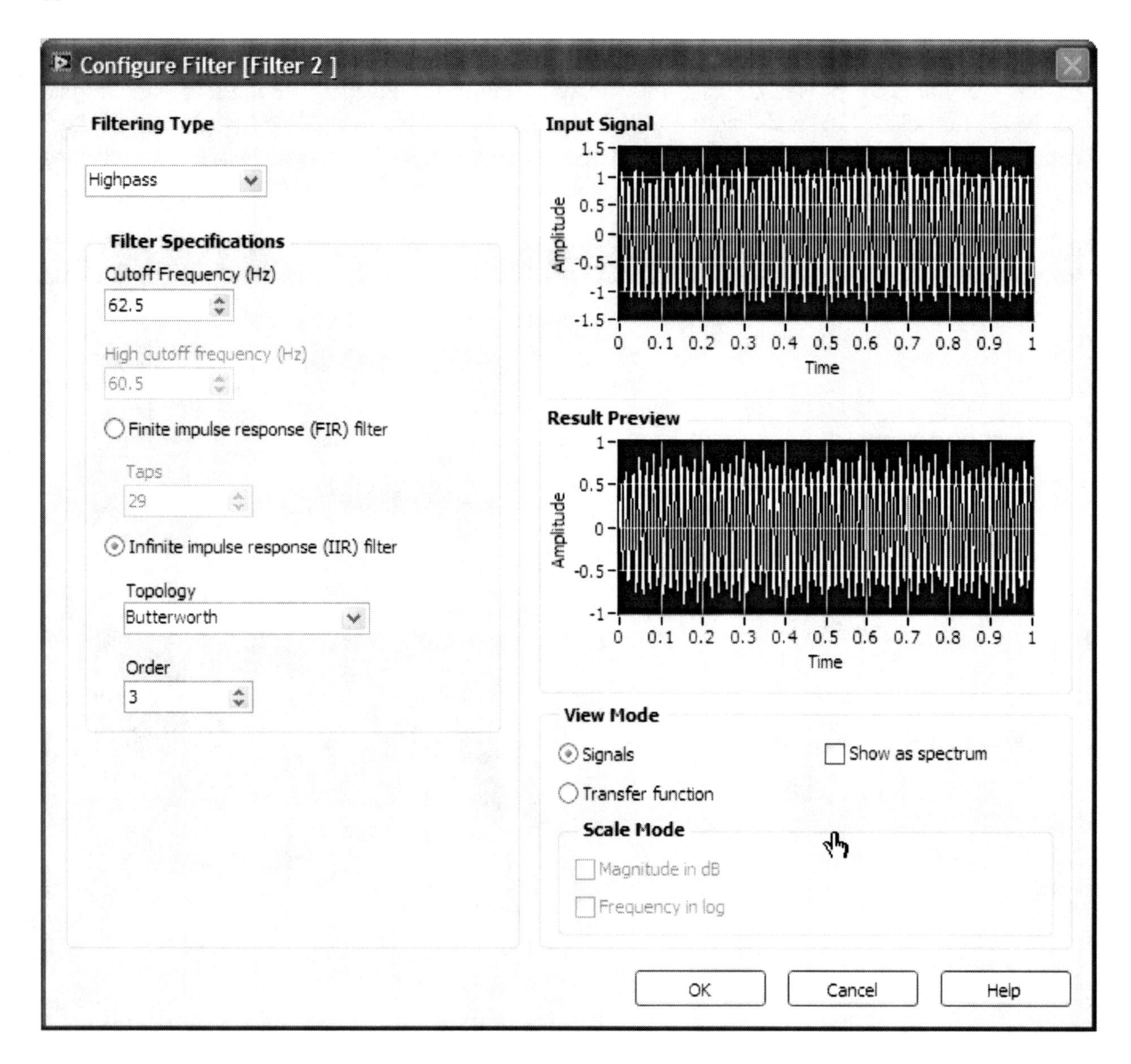

Figure 4-2-1-5 Parameters for the "Filter 2" function block. This is a high-pass filter that is configured to pass frequencies above 62.5 Hz.

4.2.2 FAST FOURIER TRANSFORM OF A SIMULATED SIGNAL

Fourier Transforms are useful for determining the frequency spectrum of non-sinusoidal periodic waveforms. LabVIEW can compute a waveform's FFT (Fast Fourier Transform) in real-time, as it is received.

Problem:

Use LabVIEW to perform a FFT (Fast Fourier Transform) on a LabVIEW "Simulated Signal". Display the "Simulated Signal" waveform and its FFT on "Front Panel" "Waveform Chart" output blocks.

LabVIEW's signal "Spectral Measurements" FFT function block is demonstrated here.

Solution:

1) Create the VI of the "Front Panel" shown in Figure 4-2-2-1 and the "Block Diagram" shown in Figure 4-2-2-2.

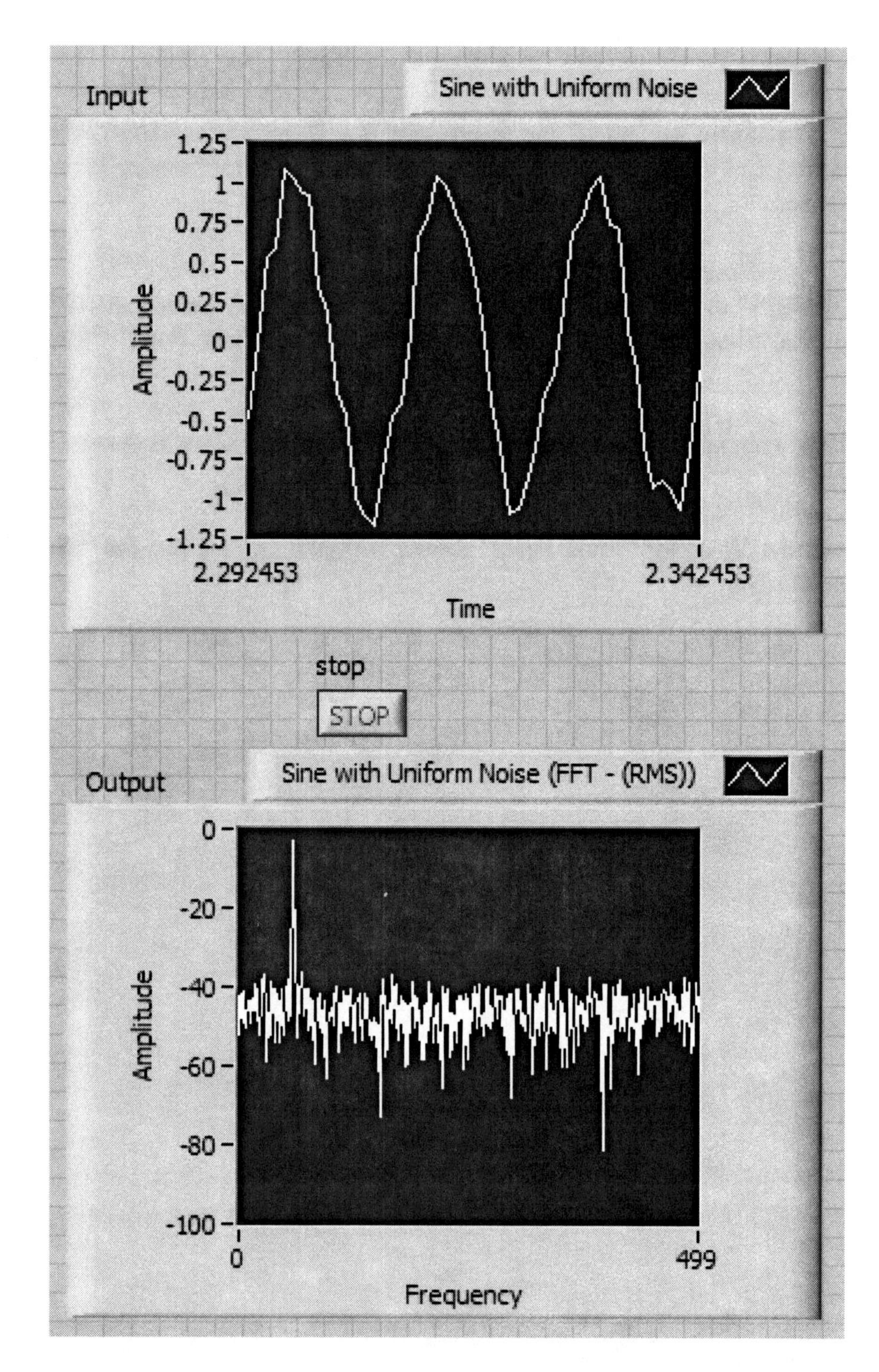

Figure 4-2-2-1 "Front Panel". "Input" shows the unfiltered signal. "Output" shows the FFT of the unfiltered signal.

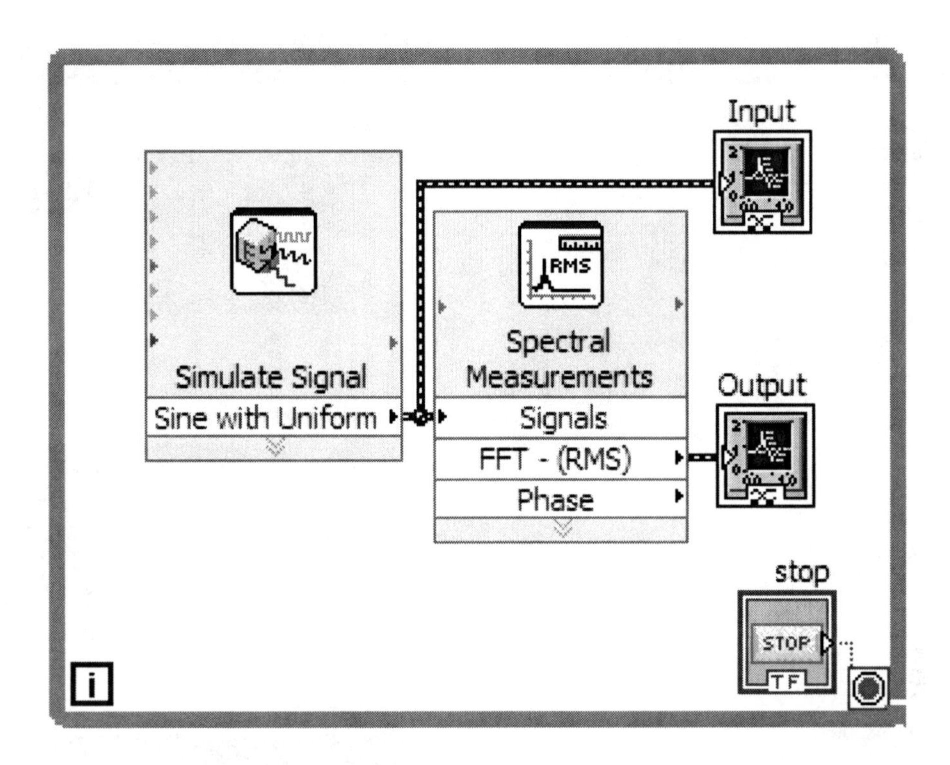

Figure 4-2-2-2 "Block Diagram". The "Simulate Signal" function block generates the signal. The "Spectral Measurements" function block creates the FFT of the signal.

2) This VI operates continuously inside a "While Loop" until one of the "Stop" buttons is left-clicked.

3) The "Simulated Signal" function block has the same parameters as in Figure 4-2-1-3 of Section 4.2.1.

4) The "Spectral Measurements" function block is placed in the "Block Diagram" by left-clicking on "View", "Functions Palette", "Express", and "Signal Analysis" Left-click, hold and drag the "Spectral Measurements" function block into the "While Loop". The "Spectral Measurements" parameter setting window appears automatically. Later, it can be selected by right-clicking on the "Spectral Measurements" function block and left-clicking on "Properties". Leave the default parameter settings. After the parameter setting window is left-clicked "OK", its "Spectral Measurements" function block will have its "Measurement Out" output automatically change to "FFT – (RMS)". See Figure 4-2-2-3.

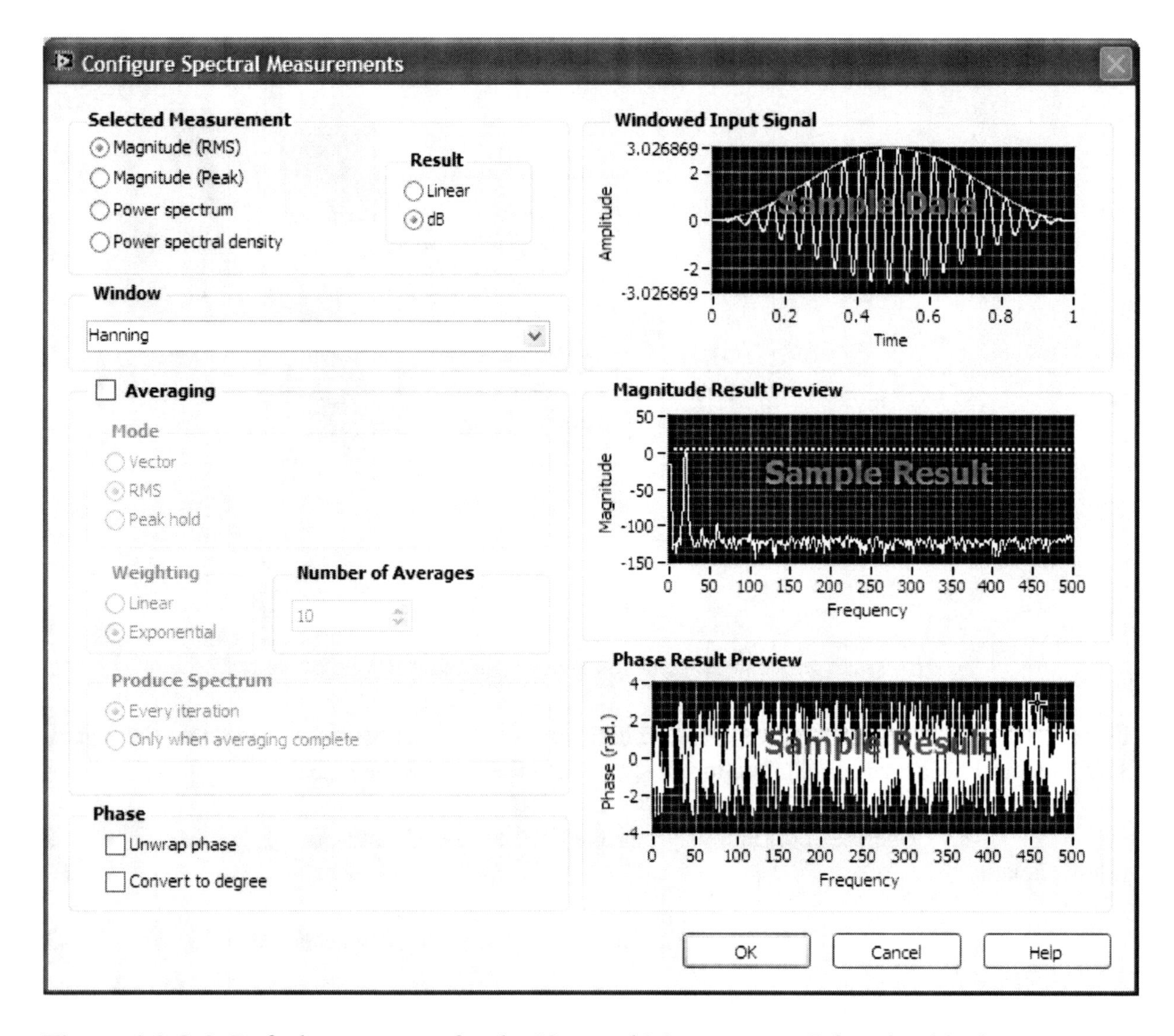

Figure 4-2-2-3 Default parameters for the "Spectral Measurements" function block.

5) The "Output" FFT seen in Figure 4-2-2-1 is plotted from 0 to 499 Hz. 0 to 499 Hz are default values that the "Output" chart selected. After the VI has been run and stopped, the "Output" chart of Figure 4-2-2-1 is displayed.

6) Portions of the "Output" chart can be selected for examination after the VI has stopped by selecting "Chart Properties:". To select "Chart Properties:", right-click on the "Output" chart, left-click on "Parameters", "Scales" and "Autoscale", to turn "Autoscale" off. Type in 40 and 80, and left-click "OK". Figure 4-2-2-4 shows the "Chart Properties:" set to display 40 to 80 Hz. Figure 4-2-2-5 shows the "Output" chart with that selection.

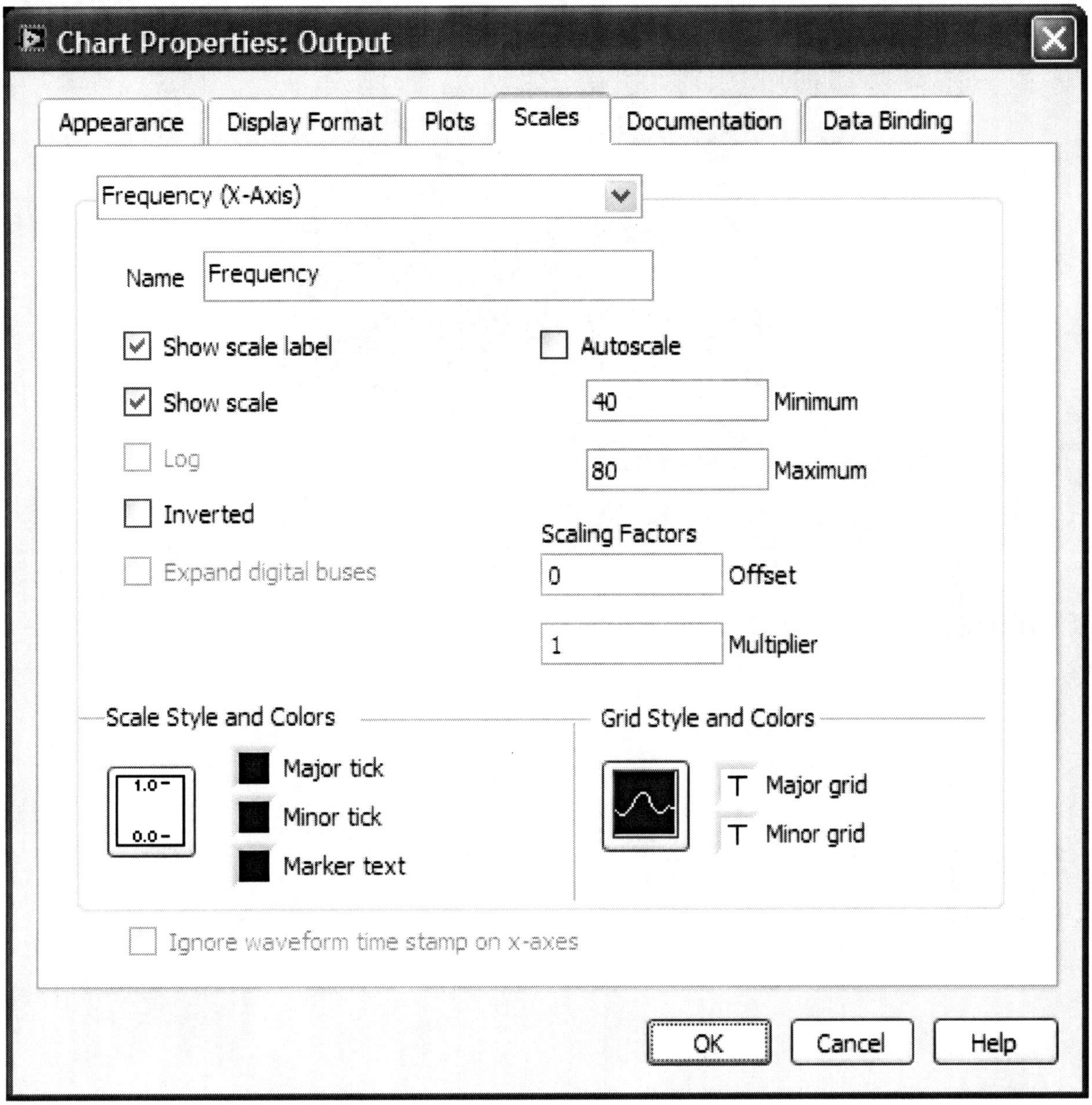

Figure 4-2-2-4 Parameters for the FFT "Output" chart with the frequency range set from 40 to 80 Hz.

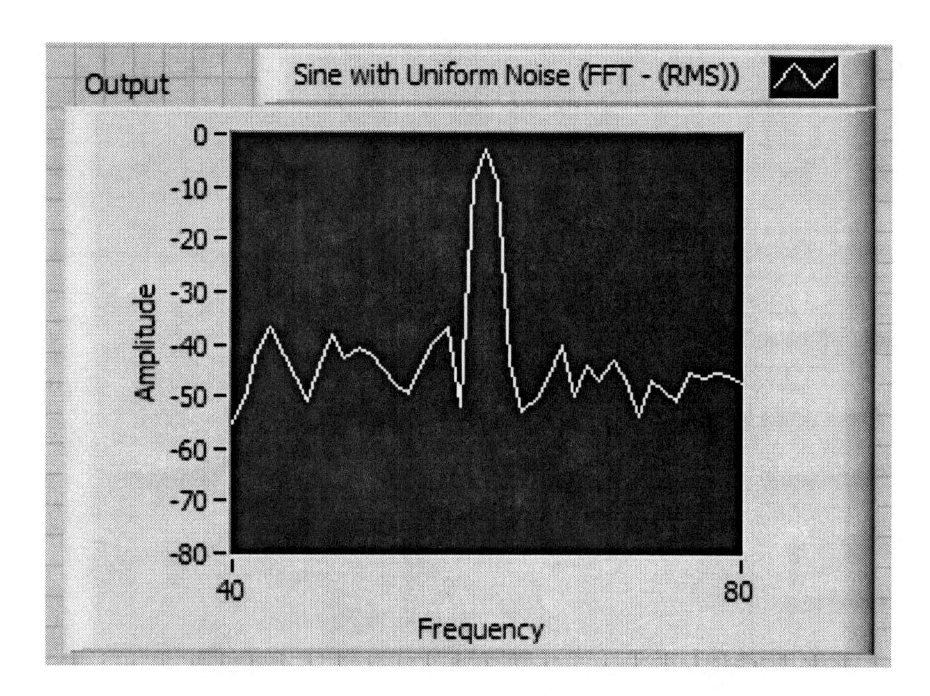

Figure 4-2-2-5 FFT "Output" chart with the frequency range set from 40 to 80 Hz.

4.2.3 STATISTICAL ANALYSIS OF A SIMULATED SIGNAL

Problem:

Use LabVIEW to determine the running average, standard deviation, maximum, and minimum of a LabVIEW simulated signal. Display these and the simulated signal on the "Front Panel".

LabVIEW's signal "Statistics" and "Merge Signals" function blocks are demonstrated here.

Solution:

1) Create the VI of the "Front Panel" shown in Figure 4-2-3-1 and the "Block Diagram" shown in Figure 4-2-3-2.

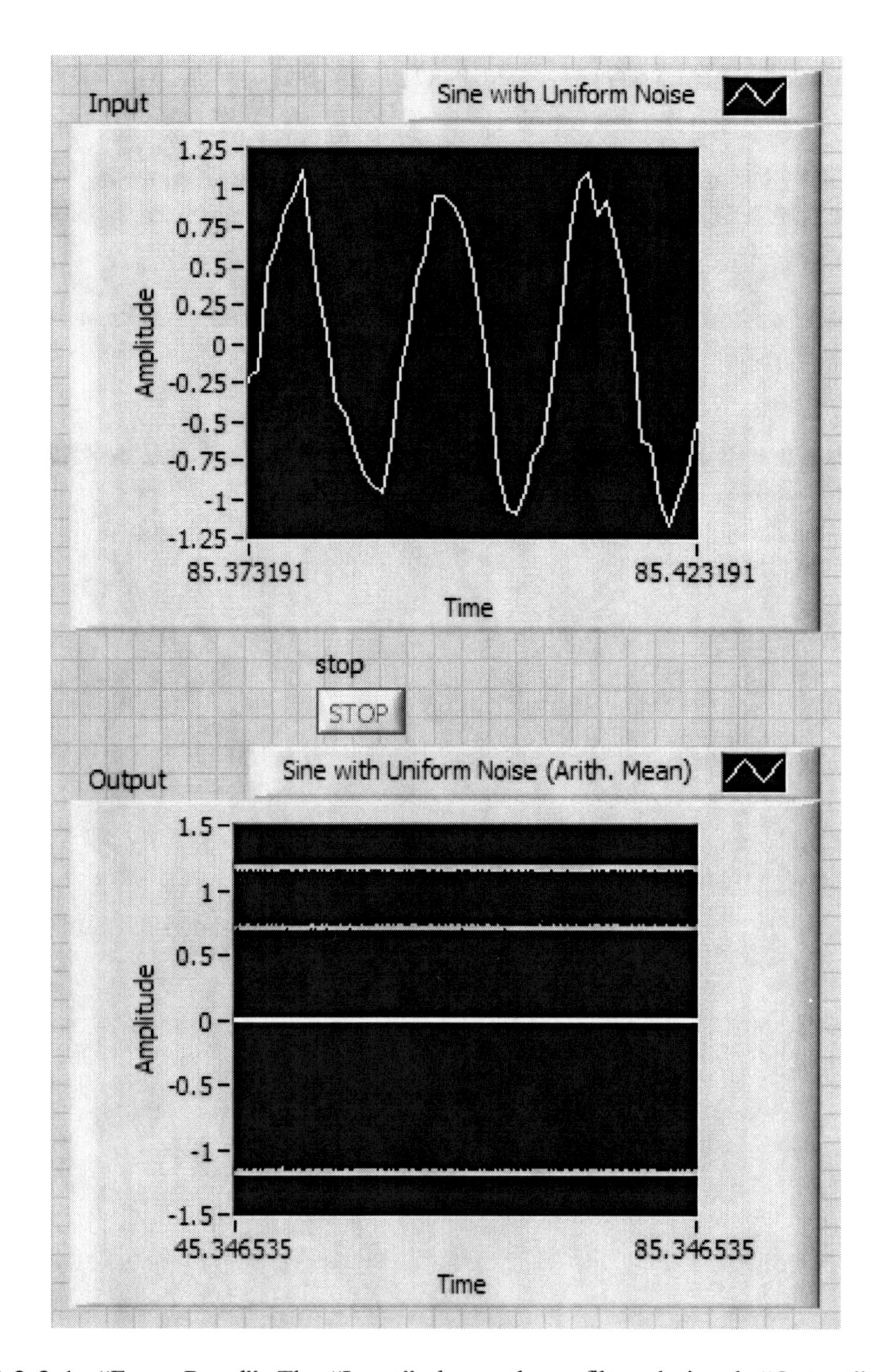

Figure 4-2-3-1 "Front Panel". The "Input" shows the unfiltered signal. "Output" shows the maximum at 1.2 (green trace), the standard deviation at .7 (red trace), the average at 0 (white trace), and the minimum at -1.2 (blue trace).

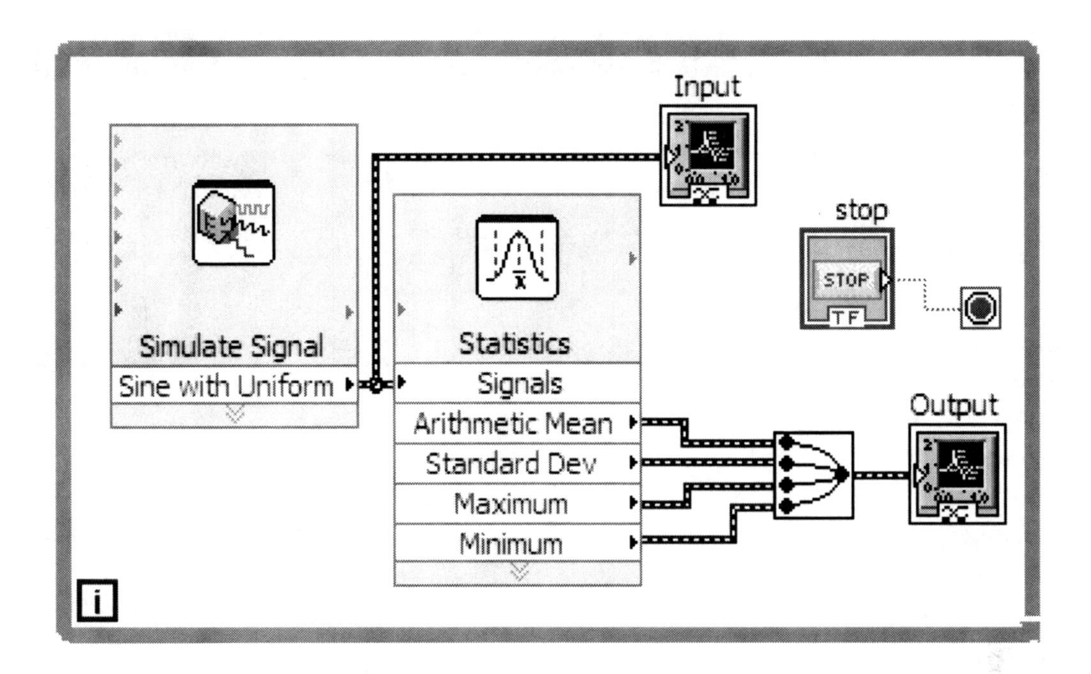

Figure 4-2-3-2 "Block Diagram". The "Simulate Signal" function block generates the signal. The "Statics" function block creates the charted statistical data.

2) This VI operates continuously inside a "While Loop" until one of the "Stop" buttons is left-clicked.

3) The "Simulated Signal" function block has the same parameters as in Figure 4-2-1-3 of Section 4.2.1.

4) The "Statistics" function block is placed in the "Block Diagram" by left-clicking on "View", "Functions Palette", "Express", and "Signal Analysis". " Left-click, hold and drag the "Statistics" function block into the "While Loop". The "Configure Statistics" parameters window opens automatically. The appropriate boxes are checked. This window can be opened later by right-clicking on the "Statistics" function block and left-clicking on "Properties". See Figure 4-2-3-3.

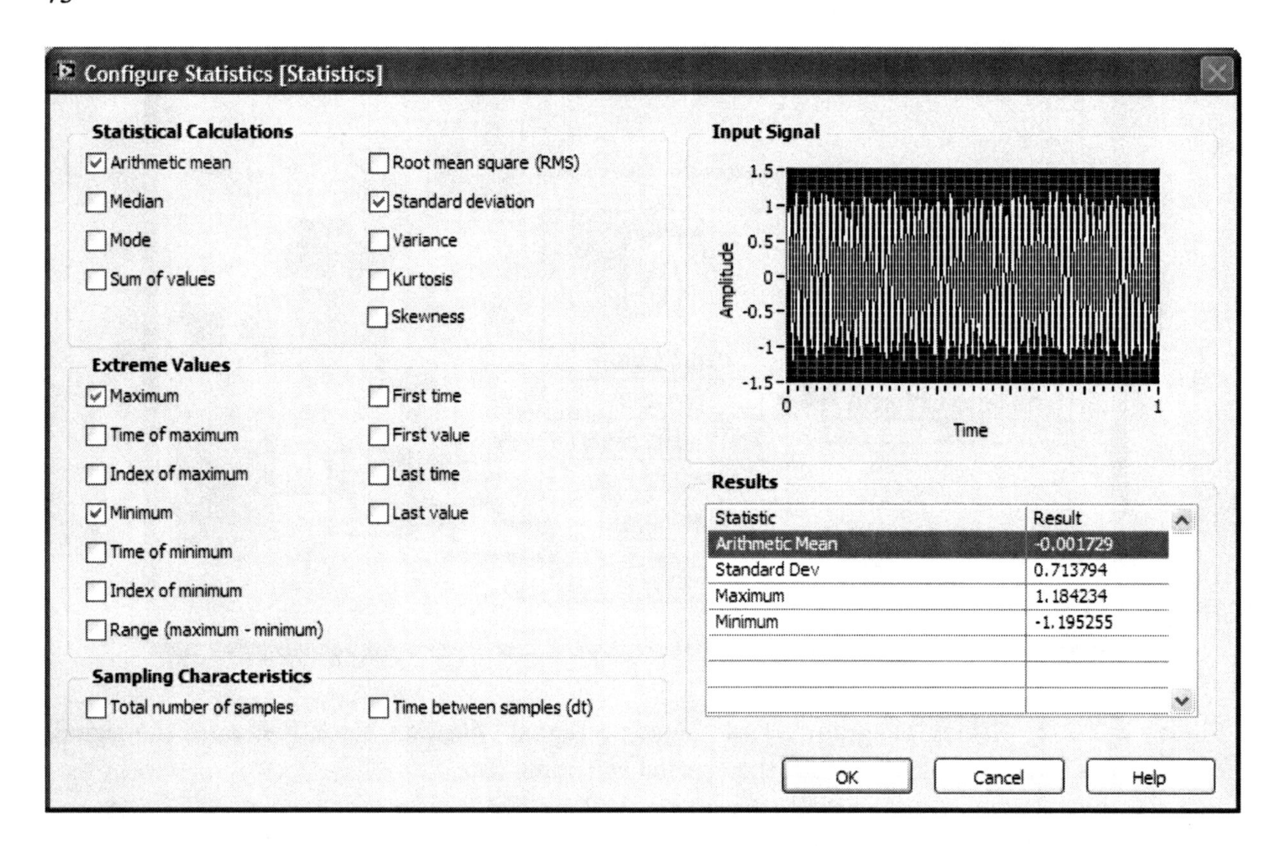

Figure 4-2-3-3 Parameters for the “Statistics” block.

4.3 SIMULATED CONNECTION OF LabVIEW TO EQUIPMENT THROUGH DAQmx SIMULATED DEVICES

With NI-DAQ and NI-DAQmx driver software, VIs can be connected to DAQ (data acquisition) devices to read sensors and control devices

National Instruments "NI-DAQmx Simulated Devices" driver software allows users to test and run VIs without being connected to actual DAQ devices. "NI-DAQmx Simulated Devices" software is free.

"NI-DAQmx Simulated Devices" driver software cannot import or export data as a physical NI-DAQ can. When LabVIEW directs a "NI-DAQmx Simulated Output Device" to output a voltage, the "NI-DAQmx Simulated Output Device" will not create a voltage that can be received by an output device or displayed by the VI. Likewise, a "NI-DAQmx Simulated Input Device" cannot receive voltage from any input device or the VI.

"NI-DAQmx Simulated Devices" driver software can create in-computer-only outputs that are useful for testing VIs.

See Section 1.6 for information on downloading NI-DAQmx drivers. Once the NI-DAQmx drivers have been downloaded, then a particular "NI-DAQmx Simulated Device" can be selected.

In LabVIEW VIs, "DAQ Assistant" function blocks are used to communicate to physical DAQs and "NI-DAQmx Simulated Devices". A "DAQ Assistant" function block and its associated parameters make up a sub-VI (subroutine) that is called a LabVIEW "Task".

4.3.1 SIMULATED STRIP CHART RECORDER CIRCUIT

Problem:

Use LabVIEW with a "NI-DAQmx USB-6210 Simulated Device" to simulate a strip chart recorder measuring two voltages in the -10 to +10 volt range.

4.3.1.1 Solution Using "Express VI" "DAQ Assistant"

LabVIEW's "NI-DAQmx USB-6210 Simulated Device", "DAQ Assistant" function block, "Run Continuously" command, and "Split Signals" function block are demonstrated here.

Solution:

1) The electrical circuit that is simulated is shown in Figure 4-3-1-1-1.

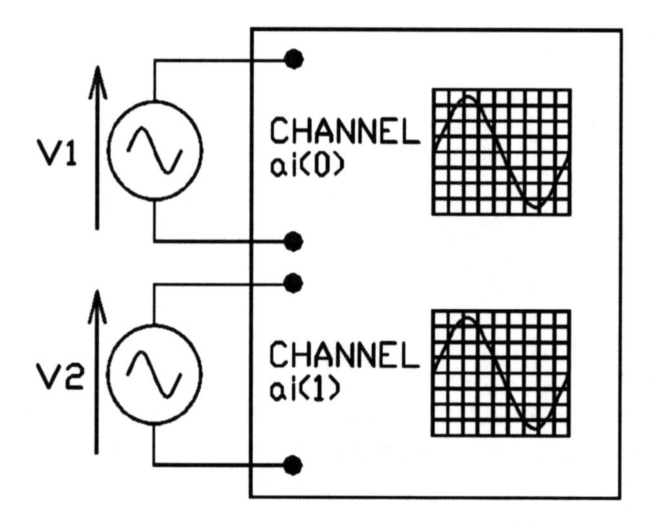

Figure 4-3-1-1-1 Strip chart recorder with two inputs.

2) To install the driver for the "NI-DAQmx USB-6210 Simulated Device":

a) If not already done, download the NI-DAQmx drivers. See Section 1.6.

b) Start the NI software to produce the "My System – Measurement & Instrumentation Explorer" window, as in Section 1.5.

c) Right-click on "Devices and Interfaces" and left-click on "Create New".

d) In the appearing window left-double-click on "Simulated NI-DAQmx Device or Modular Instrument".

e) Find the USB-6210 by typing it into the "Search" location at the top of the "Create Simulated NI-DAQmx Device" window.

f) Left-click on "USB-6210" and "OK". Now USB-6210 appears in the "Devices and Interfaces" folder. Verify this by opening the folder.

g) Note that the icon for USB-6210 is yellow. "Simulated" NI-DAQmx device icons are yellow. Physical NI-DAQmx device icons are green.

3) Launch LabVIEW.

4) The "DAQ Assistant" function block is placed in the "Block Diagram" by left-clicking on "View", "Functions Palette", "Express", and "Input". Left-click, hold and drag the "DAQ Assistant" onto the "Block Diagram".

5) In the window that comes up left-click on "Acquire Signals", "Analog Input", "Voltage", "ai0", and "Finish".

6) A "DAQ Assistant" properties window appears with its "Express Task" tab open. Leave the default values. Later, if a change of values is desired, right-click on the "DAQ Assistant" function block and left-click on "Properties" to get this window.

7) When the "DAQ Assistant" window "Run" is left-clicked a graph of one cycle of the voltage generated by the "NI-DAQmx USB-6210 Simulated Device" is displayed. One cycle is equal to 100 data points with .001 second per data point. The .001 second per data point is determined by the 1k Hz rate. See Figure 4-3-1-1-2.

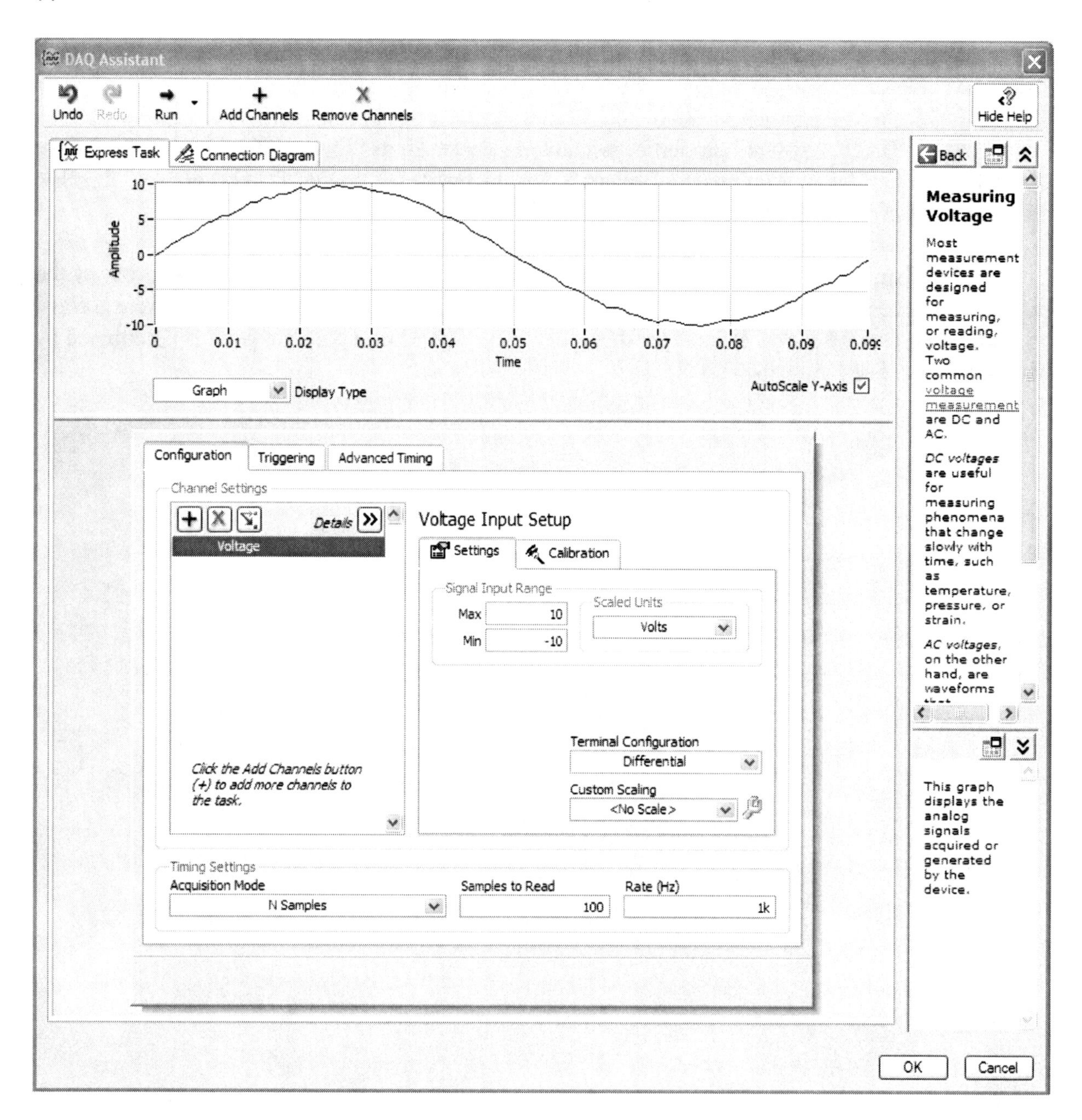

Figure 4-3-1-1-2 "DAQ Assistant" window, "Express Task" tab, after the "DAQ Assistant" window's "Run" is left-clicked.

8) Left-click "OK" on the "DAQ Assistant" properties window.

9) Left-click, hold, and drag down the bottom of the "DAQ Assistant" function block so that its inputs and outputs are visible.

10) Put a "Waveform Chart" output block on the "Front Panel" by left-clicking on "View", "Controls Palette", "Express", "Output", and "Graph Indicators". Left-click, hold and drag the "Waveform Chart" onto the "Front Panel".

11) Right-click on the "Front Panel" "Waveform Chart" and left-click on "Properties". In the appearing "Chart Properties: Waveform Chart" left-click on the "Display Format" tab and "Floating Point". Set "Digits" to 4. Then left-click on the "Scales" tab, left-click the "Autoscale" off, and set the "Minimum" to 0 and the "Maximum" to .1. Leave the other default values and left-click "OK".

12) Connect a dataflow wire from the "DAQ Assistant" "data" output to the "Waveform Chart" input.

13) Left-click the "Run Continuously" command. It is next to the main toolbar's "Run" command. See the results in Figures 4-3-1-1-3 and 4-3-1-1-4.

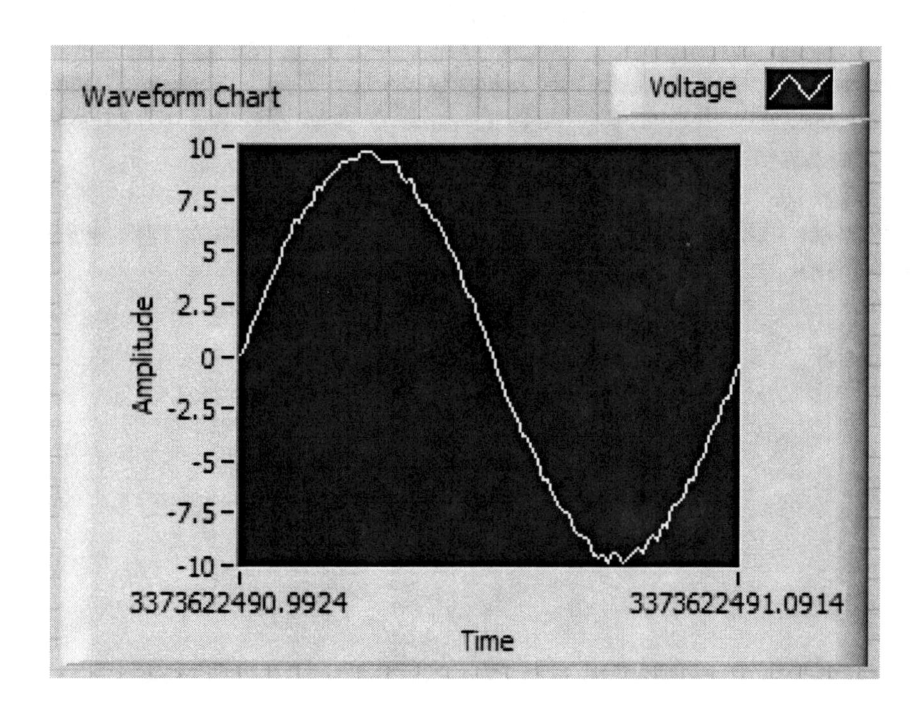

Figure 4-3-1-1-3 "Front Panel" with one signal going to the one "Waveform Chart" output block. The "Time" scale shows the seconds after 7 pm Dec. 31, 1903.

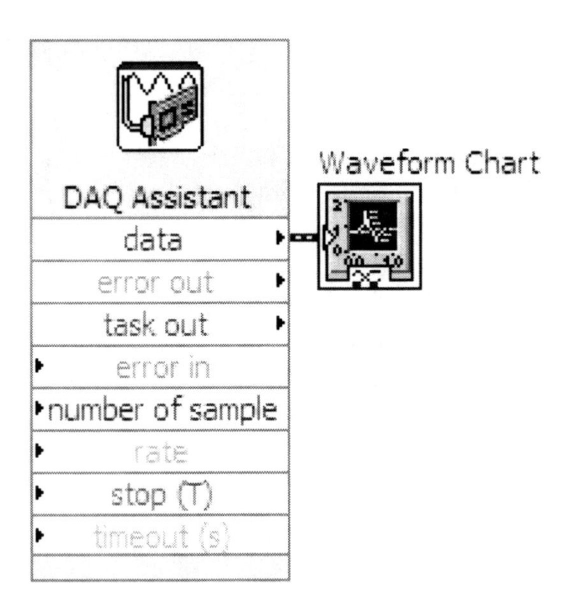

Figure 4-3-1-1-4 "Block Diagram" with the "DAQ Assistant" function block representing the "NI-DAQmx USB-6210 Simulated Device".

14) The second signal is added by bringing up the "DAQ Assistant" window of Figure 4-3-1-1-2 again and left-clicking on "Add Channels" near the top of the window. In the appearing window left-click on "Voltage", "ai(1)", and "OK".

15) When the "DAQ Assistant" window "Run" is left-clicked, two sine waves appear. See Figure 4-3-1-1-5.

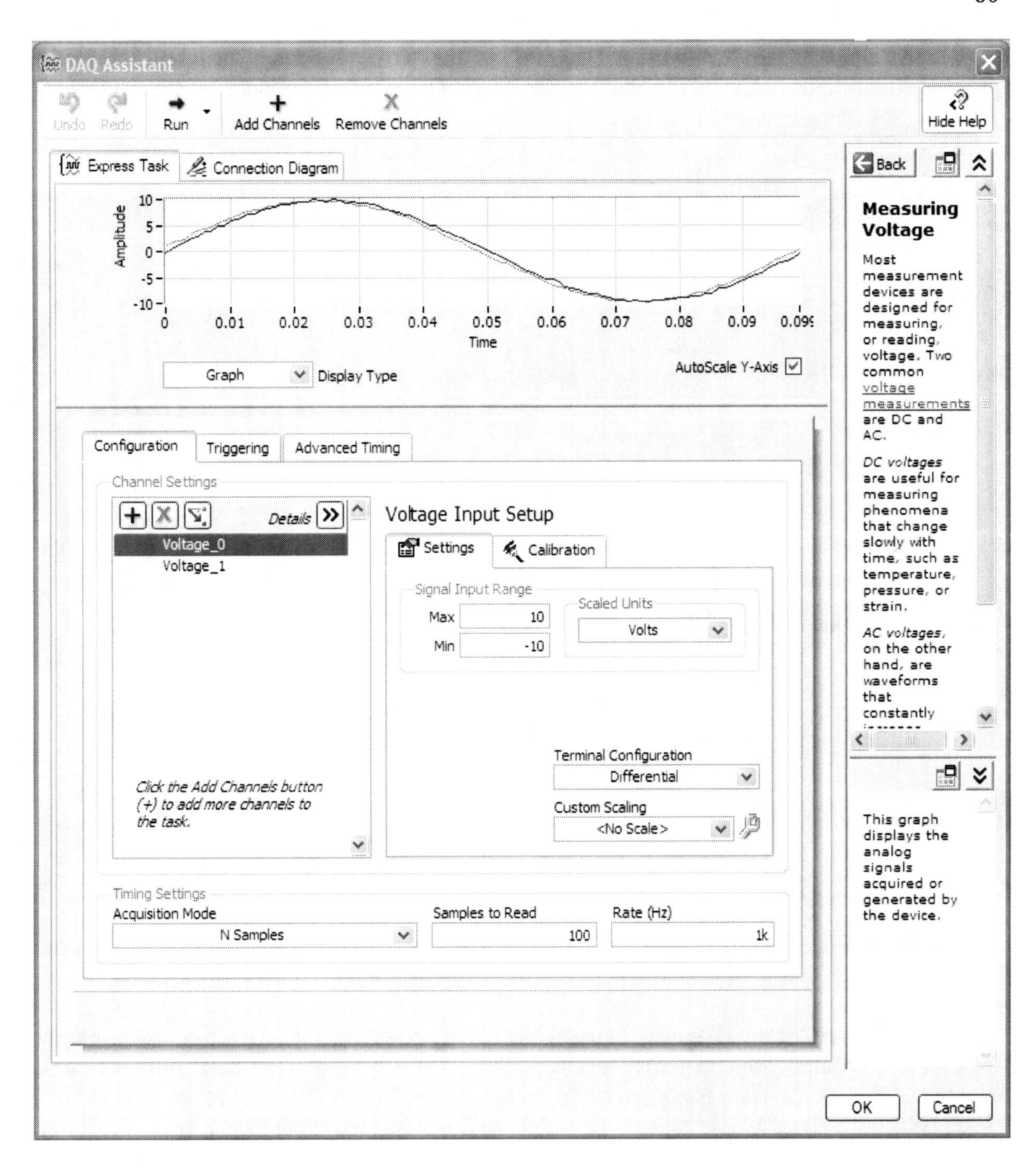

Figure 4-3-1-1-5 "DAQ Assistant" window with two channels.

16) Opening the "Connection Diagram" in the "DAQ Assistant" window shows the physical connections that would be made to the NI-DAQ USB-6210 for input ai(0). See Figure 4-3-1-1-6. The connections for ai(1) could be seen by highlighting "Voltage_1" rather than "Voltage_0".

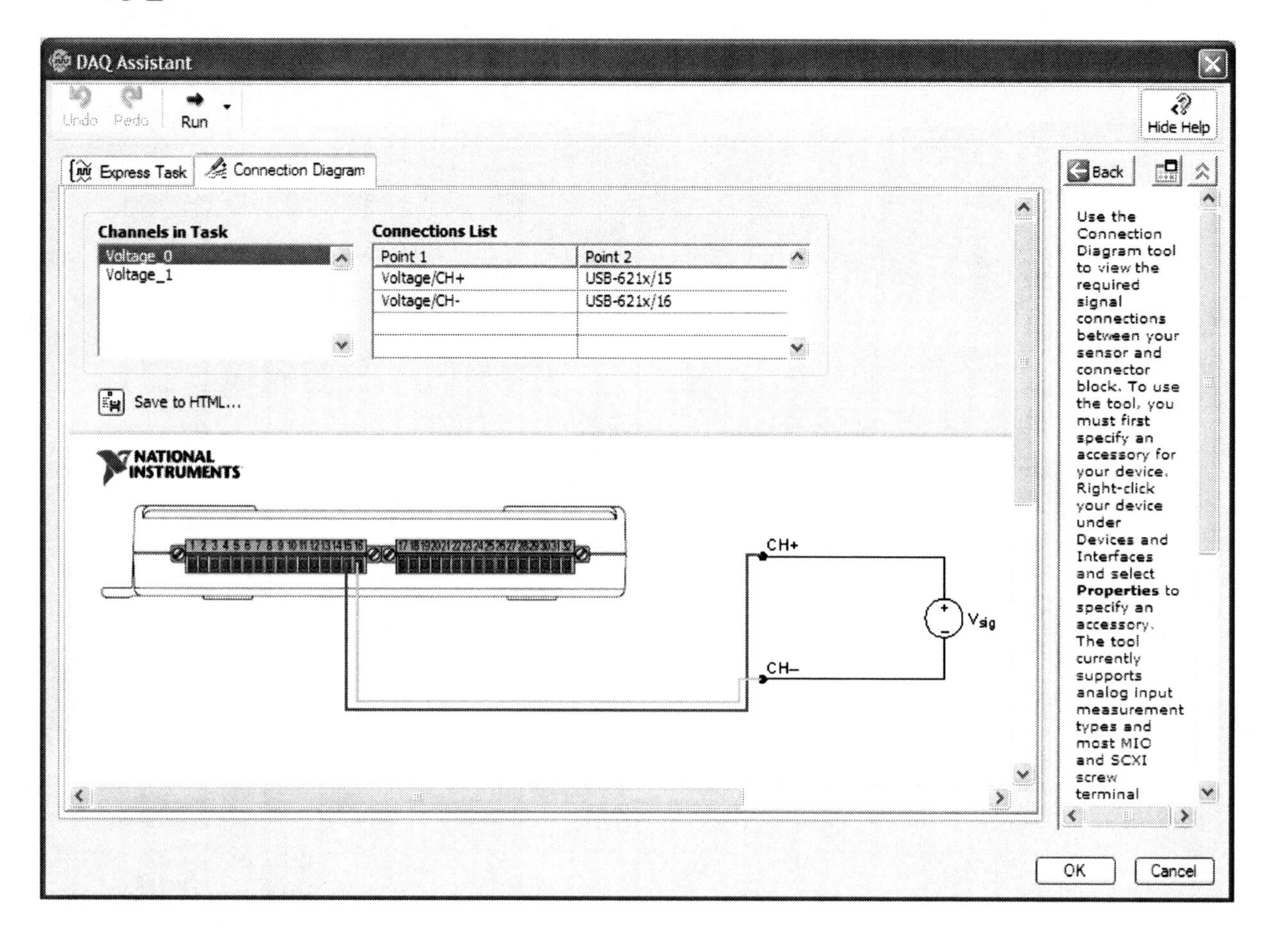

Figure 4-3-1-1-6 "DAQ Assistant" window showing the "Connection Diagram" for channel "ai(0)".

17) In the "Block Diagram" connect the "DAQ Assistant" function block data output through a two output "Split Signals" function block to two "Waveform Chart" output blocks. The "Split Signals" function block can be found by left-clicking on "View", "Functions Palette", "Express", and "Signal Manipulation". Left-click, hold and drag the "Split Signals" onto the "Front Panel". The "Block Diagram" and "Front Panel" can be seen in Figure 4-3-1-1-7 and 4-3-1-1-8.

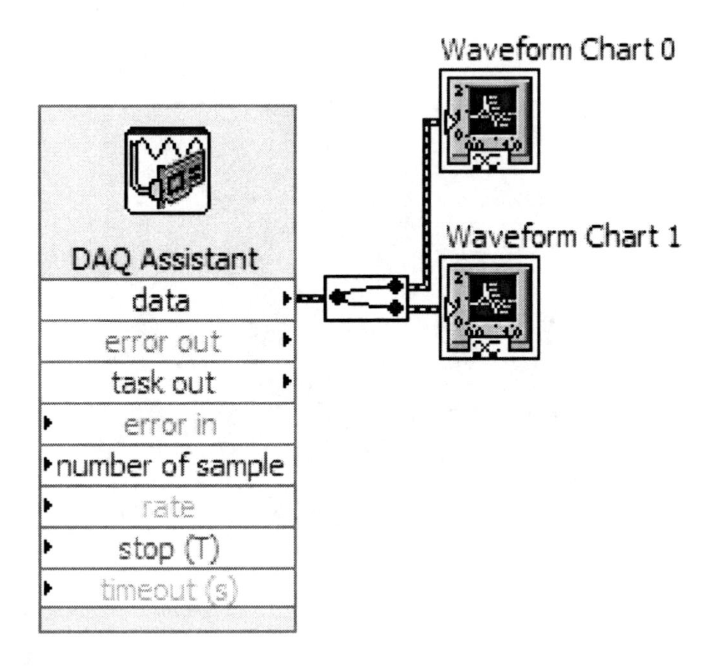

Figure 4-3-1-1-7 “Block Diagram” with two signals going to two “Waveform Chart” output blocks. The “DAQ Assistant” function block represents the “NI-DAQmx USB-6210 Simulated Device”.

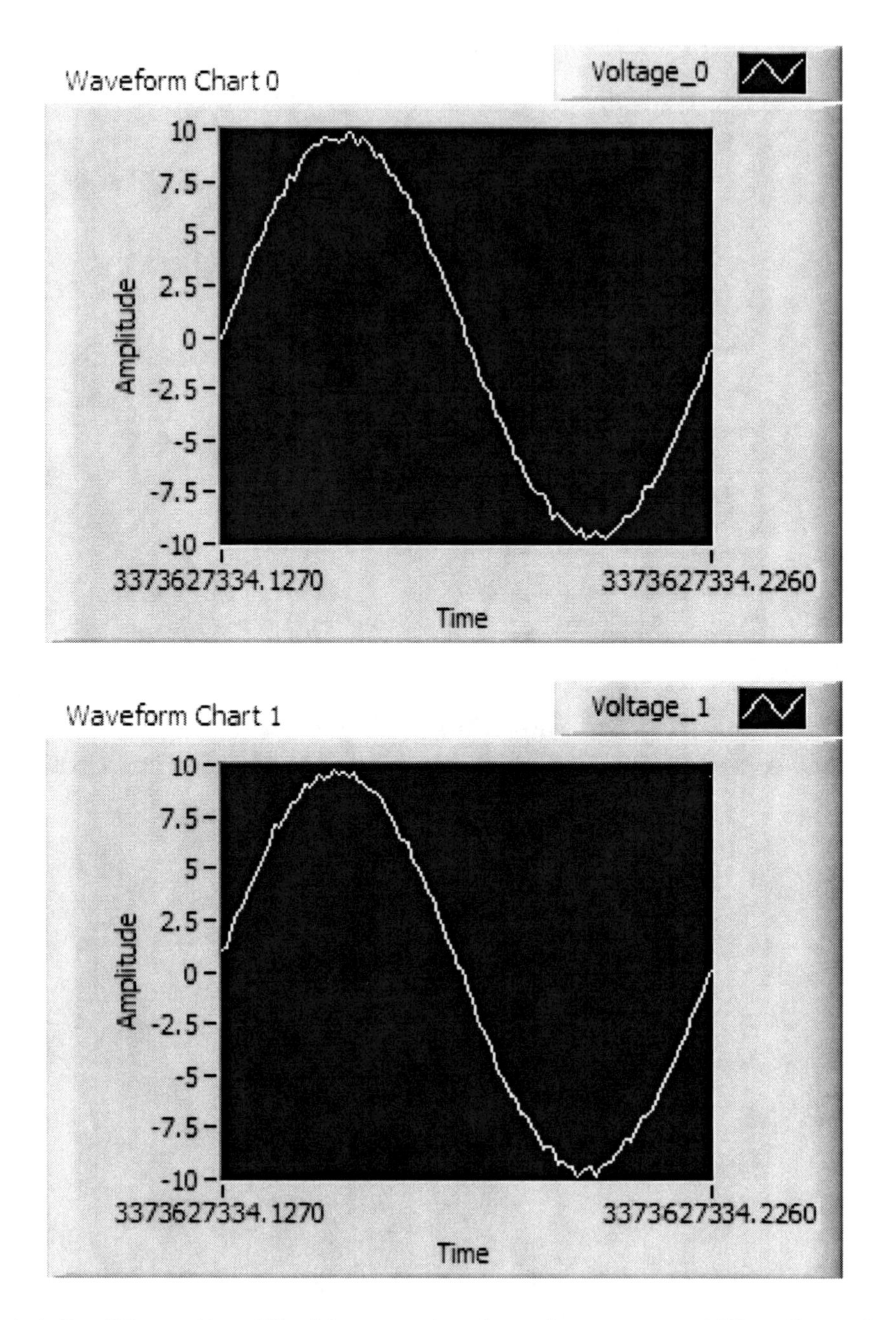

Figure 4-3-1-1-8 "Front Panel" with two signals going to two "Waveform Chart" output blocks. The "Time" scale shows the seconds after 7 pm Dec. 31, 1903.

4.3.1.2 Solution Using "Standard VI"s

This is the same problem as in 4.3.1.1 but with the more versatile "Standard VI"s rather than the easier to use "Express" "DAQ Assistant".

LabVIEW's ability to convert "Express" "DAQ Assistant" VIs to "Standard VI"s is demonstrated here.

Solution:

1) On the "Block Diagram" of Figure 4-3-1-1-7 right-click on the "DAQ Assistant" function block. Then left-click on "Generate NI-DAQmx Code". The resulting "Block Diagram" is shown in Figure 4-3-1-2-1. The resulting "Front Panel" is shown in Figure 4-3-1-2-2.

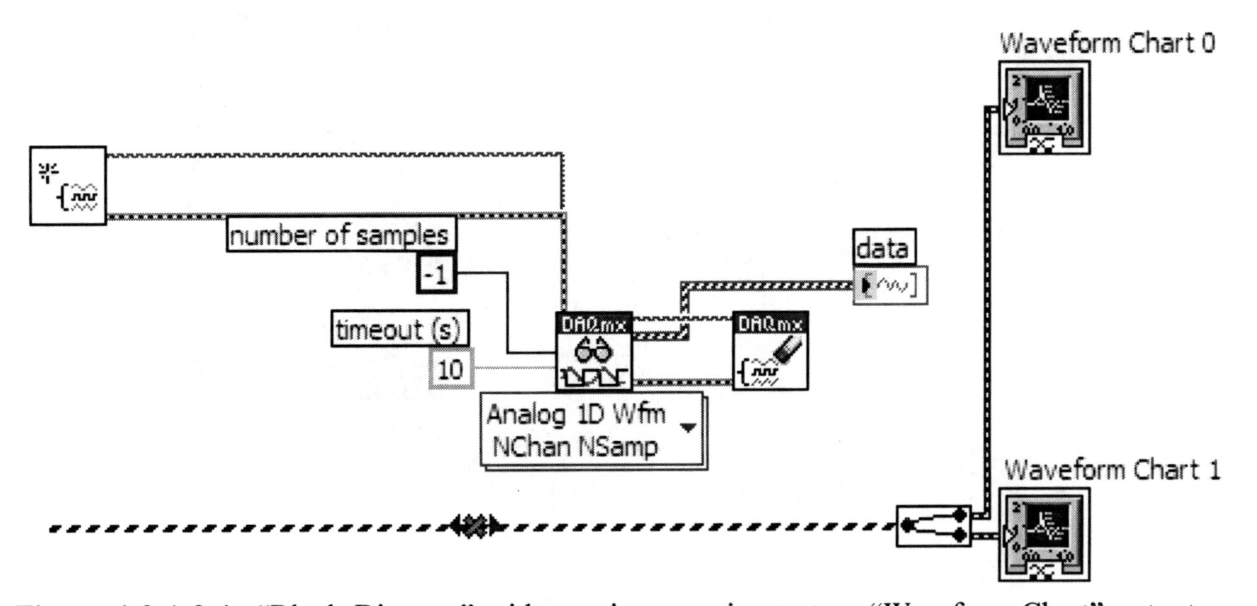

Figure 4-3-1-2-1 "Block Diagram" with two inputs going to two "Waveform Chart" output blocks. Rather than a "DAQ Assistant", "Standard VI"s are used. As before the diagram uses a "NI-DAQmx USB-6210 Simulated Device".

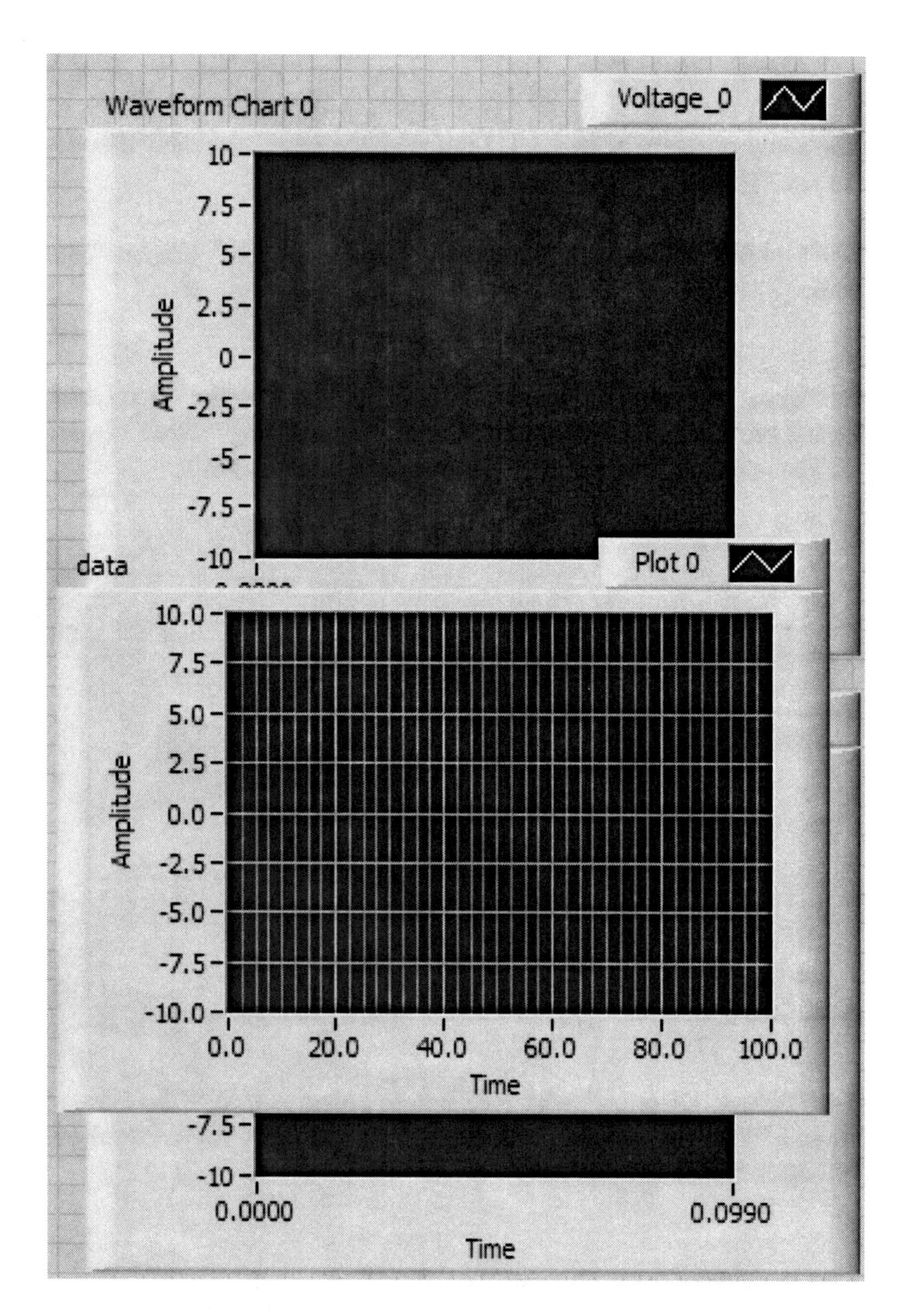

Figure 4-3-1-2-2 "Front Panel" after the "DAQ Assistant" function block is converted to "Standard VI". Note the conversion added its own "data" "Waveform Chart" output block.

2) In the "Block Diagram" of Figure 4-3-1-2-1 the "Waveform Chart 0" and "Waveform Chart 1" did not automatically connect to the "Standard VI". Erase the disconnected dataflow wires and then connect a new dataflow wire from the "Split Signals" function block to the "DAQmx" data output. This is shown in Figure 4-3-1-2-3.

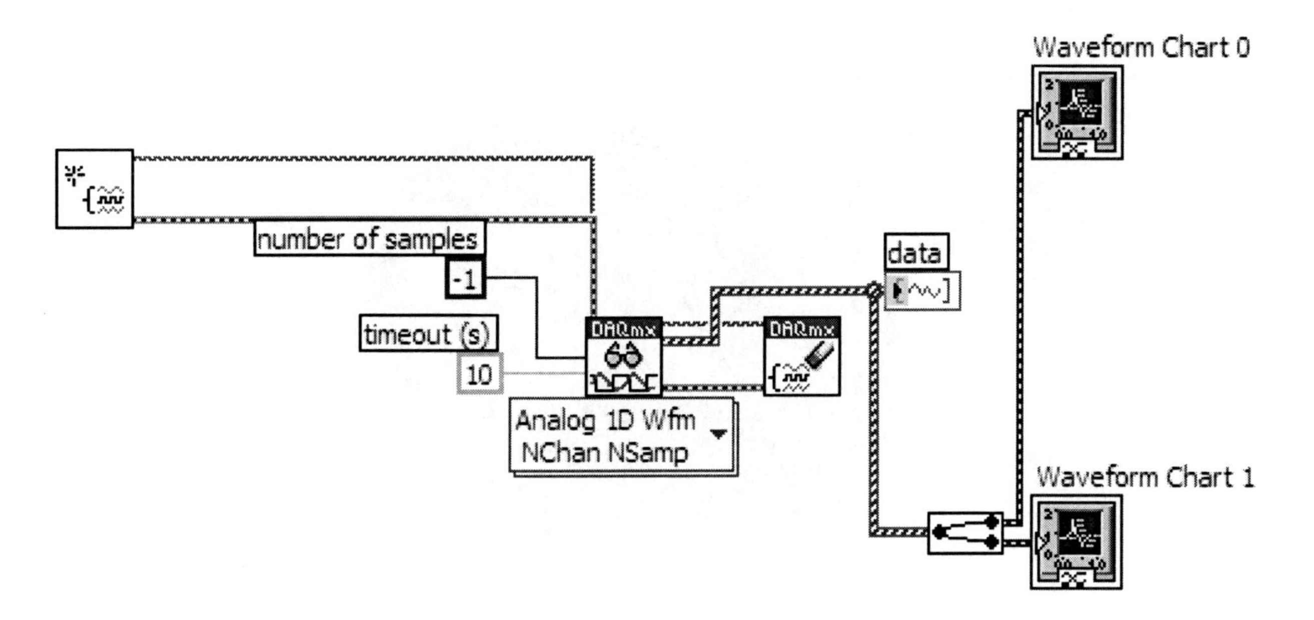

Figure 4-3-1-2-3 Corrected “Block Diagram” with two inputs going to two “Waveform Chart” output blocks.

3) The “data” “Waveform Chart” output block that was created when the “DAQ Assistant” function block was created is moved down so that all the “Waveform Chart” output blocks are visible. See the results of running the VI on the “Waveform Chart” output blocks in Figure 4-3-1-2-4.

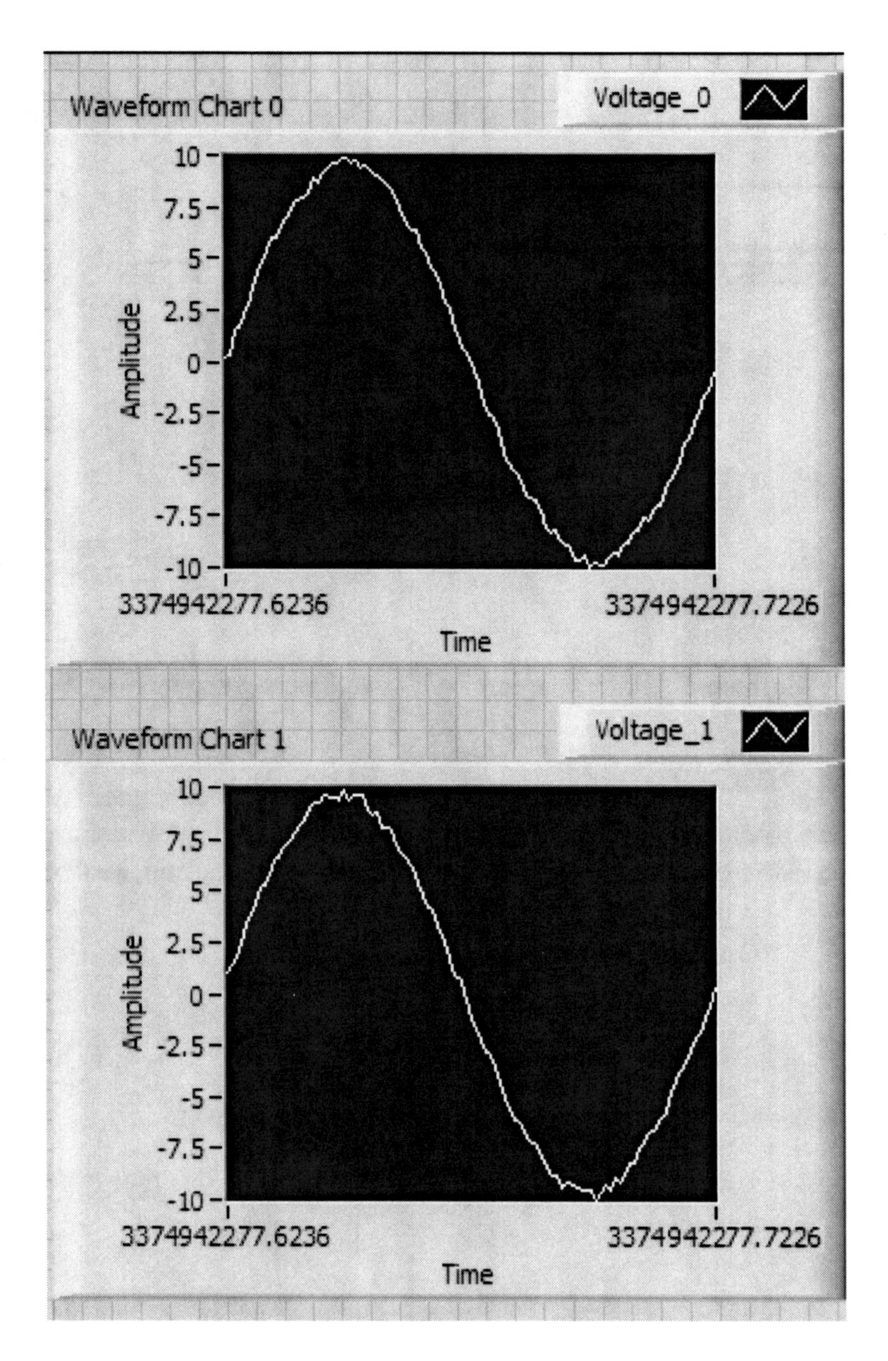

Waveform Chart 0
Voltage_0
10
7.5
5
2.5
0
-2.5
-5
-7.5
-10
Amplitude
3374942277.6236
3374942277.7226
Time
Waveform Chart 1
Voltage_1
10
7.5
5
2.5
0
-2.5
-5
-7.5
-10
Amplitude
3374942277.6236
3374942277.7226
Time

Continued

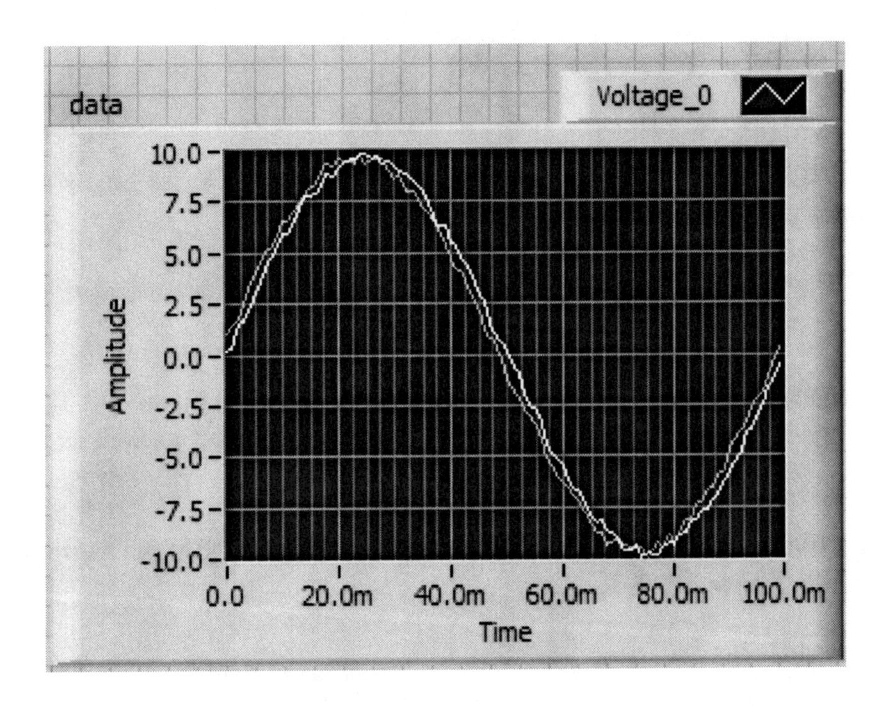

Figure 4-3-1-2-4 "Front Panel" after the "DAQ Assistant" function block is converted to a "Standard VI".

4.3.2 WRITING SIMULATED DATA TO A FILE

Problem:

Use LabVIEW with a "NI-DAQmx USB-6210 Simulated Device" to write the simulated data of the example problem of Section 4.3.1 to a file.

LabVIEW's "Write to Measurement File" output block is demonstrated here.

Solution:

1) The simulated electrical circuit is the same as that shown in Figure 4-3-1-1-1. However in this VI the numerical values of voltages V1 and V2 will be recorded as well as their waveforms graphed.

2) Add a "Write to Measurement File" output block to the "Block Diagram" of Figure 4-3-1-1-7 as shown in Figure 4-3-2-1.

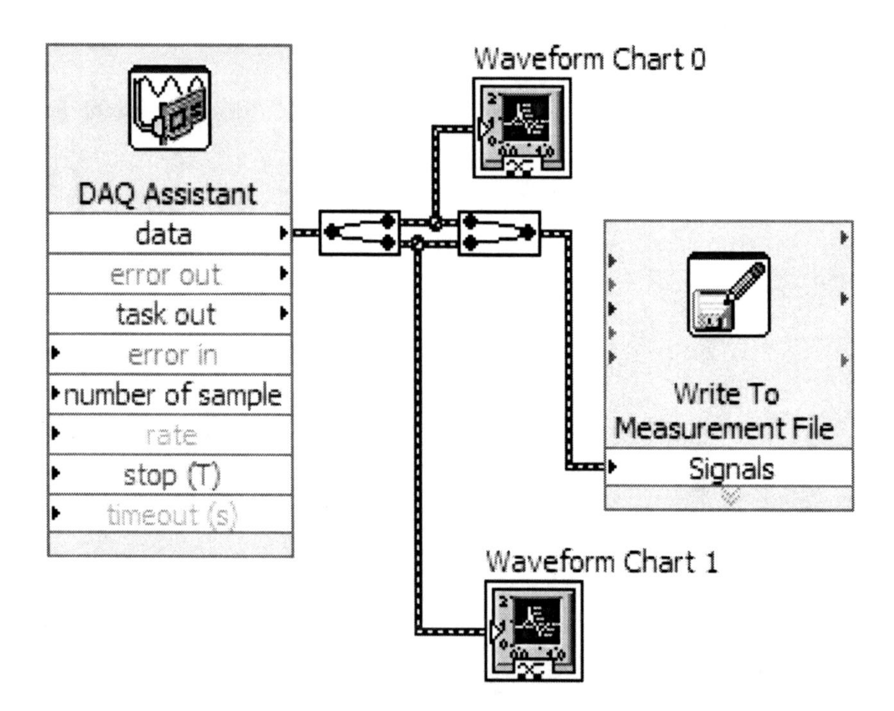

Figure 4-3-2-1 "Block Diagram" with two signals going to two "Waveform Chart" output blocks and one "Write Measurement File" output block.

3) The "Write to Measurement File" output block can be found by left-clicking on "View", "Functions Palette", "Express", and "Output". Left-click, hold and drag the "Write Measurement File" onto the "Front Panel". When it is first called up the configuration window shown in Figure 4-3-2-2 appears. Note that a file name, file description, and a check indicating "One column only" for time have been entered.

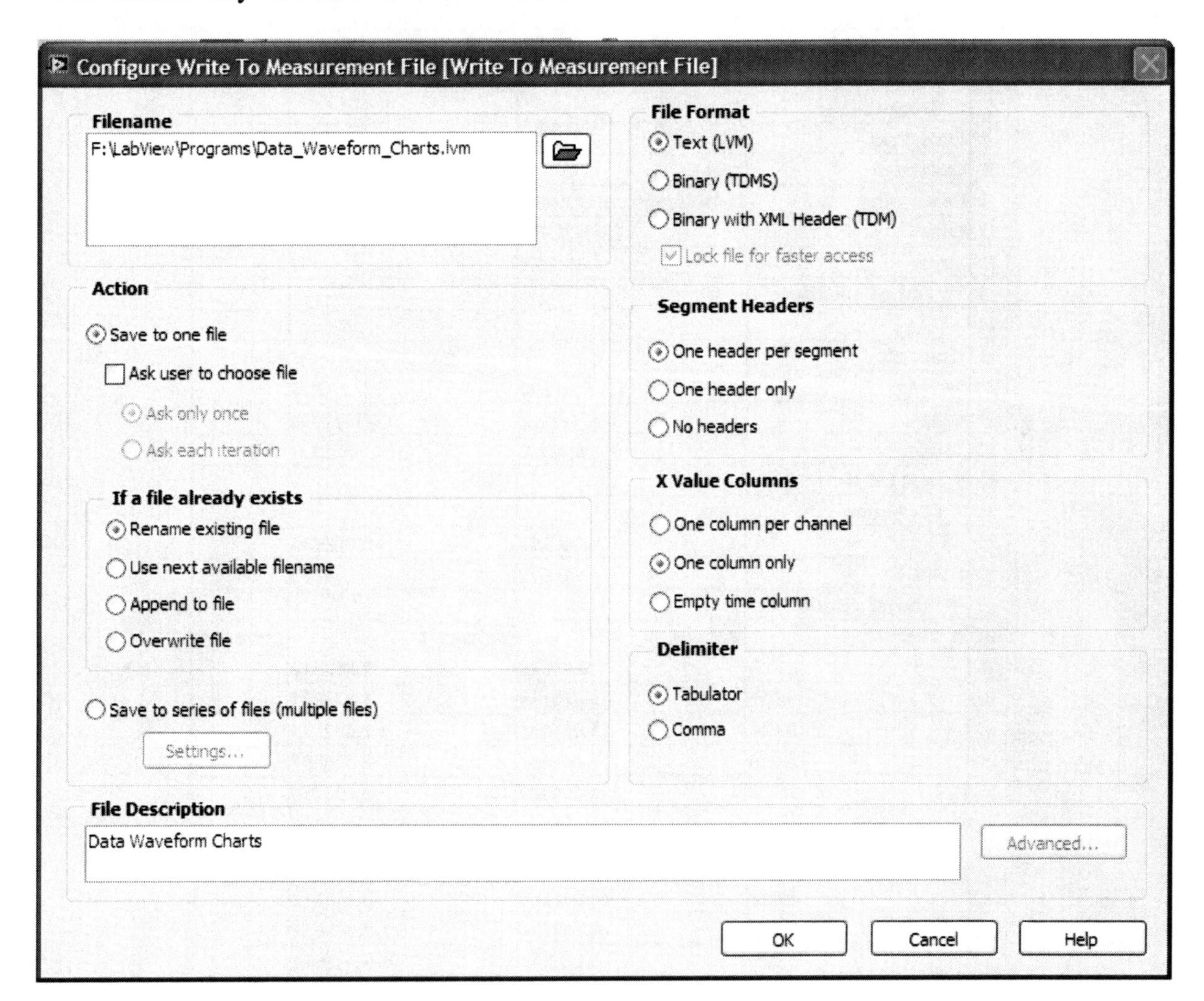

Figure 4-3-2-2 "Configure Write To Measurement File".

4) When the VI is run the data file appears at
"F:\LabView\Programs\Data_Waveform_Charts.lvm"
This file can be opened with Microsoft Excel. Figure 4-3-2-3 shows it in Excel after the columns have been widened.

Data_Waveform_Charts.lvm

	A	B	C	D
1	LabVIEW Measurement			
2	Writer_Version	2		
3	Reader_Version	2		
4	Separator	Tab		
5	Decimal_Separator	.		
6	Multi_Headings	Yes		
7	X_Columns	One		
8	Time_Pref	Absolute		
9	Operator	tubbs		
10	Description	Data Waveform Charts		
11	Date	12/19/2010		
12	Time	45:20.9		
13	***End_of_Header***			
14				
15	Channels	2		
16	Samples	100	100	
17	Date	12/19/2010	12/19/2010	
18	Time	45:21.0	45:21.0	
19	Y_Unit_Label	Volts	Volts	
20	X_Dimension	Time	Time	
21	X0	0.00E+00	0.00E+00	
22	Delta_X	0.001	0.001	
23	***End_of_Header***			
24	X_Value	Voltage_0	Voltage_1	Comment
25	0	0.04883	0.968352	
26	0.001	0.425733	1.322977	
27	0.002	1.155431	1.852168	
28	0.003	1.635487	2.706687	
29	0.004	2.636189	3.016755	
112	0.087	-6.840114	-6.482131	
113	0.088	-6.843776	-5.847652	
114	0.089	-6.188543	-5.338603	
115	0.09	-5.560778	-4.79751	
116	0.091	-4.97055	-4.379406	
117	0.092	-4.737693	-4.055605	
118	0.093	-4.401379	-3.259377	
119	0.094	-3.861507	-2.578204	
120	0.095	-2.735069	-2.026124	
121	0.096	-2.136601	-1.684011	
122	0.097	-1.597339	-1.239662	
123	0.098	-1.356853	-0.566424	
124	0.099	-0.477309	0.172124	

Figure 4-3-2-3 "Write To Measurement File" data output displayed in Microsoft Excel.

4.3.3 SIMULATED THERMOSTAT CIRCUIT

Problem:

Use LabVIEW with a "NI-DAQmx USB-6210 Simulated Device" to simulate a thermocouple controlled furnace. There should be a 5 second delay before a turn-on or turn-off signal is sent to the furnace heater contactor. The circuit diagram is shown in Figure 4-3-3-1.

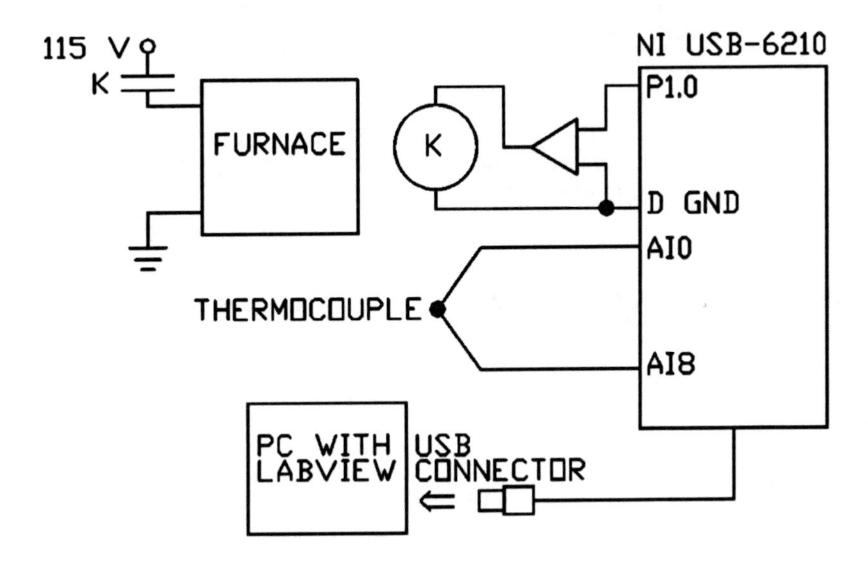

Figure 4-3-3-1 Furnace control circuit.

The new functions demonstrated here are NI USB-6210's inputting an analog signal, NI USB-6210's outputting a digital signal, "Round LED" output block, "Less?" function block, "Convert from Dynamic Data" function block, "Build Array" function block, and "Wait" function block.

Solution:

1) In Figure 4-3-3-1 notice the isolation amplifier between the NI USB-6210 digital output and the contactor coil, K. This is needed because the NI USB-6210 can create only a relatively low output current, not enough to power a contactor coil.

2) Create the VI of the "Front Panel" of Figure 4-3-3-2 and the "Block Diagram" of Figure 4-3-3-3.

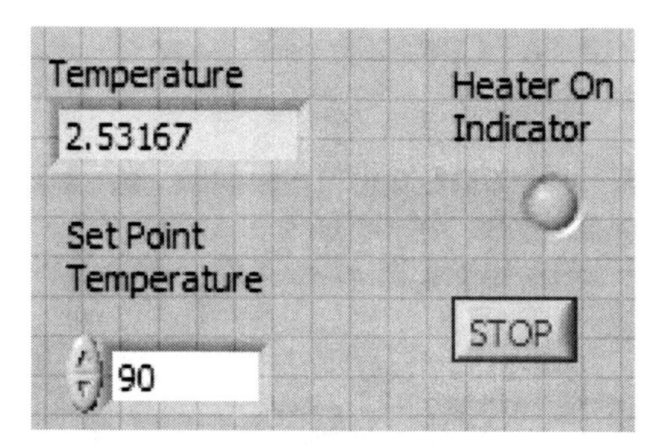

Figure 4-3-3-2 "Front Panel" displaying the set point temperature in °C, thermocouple output converted to °C, and "Heater On Indicator".

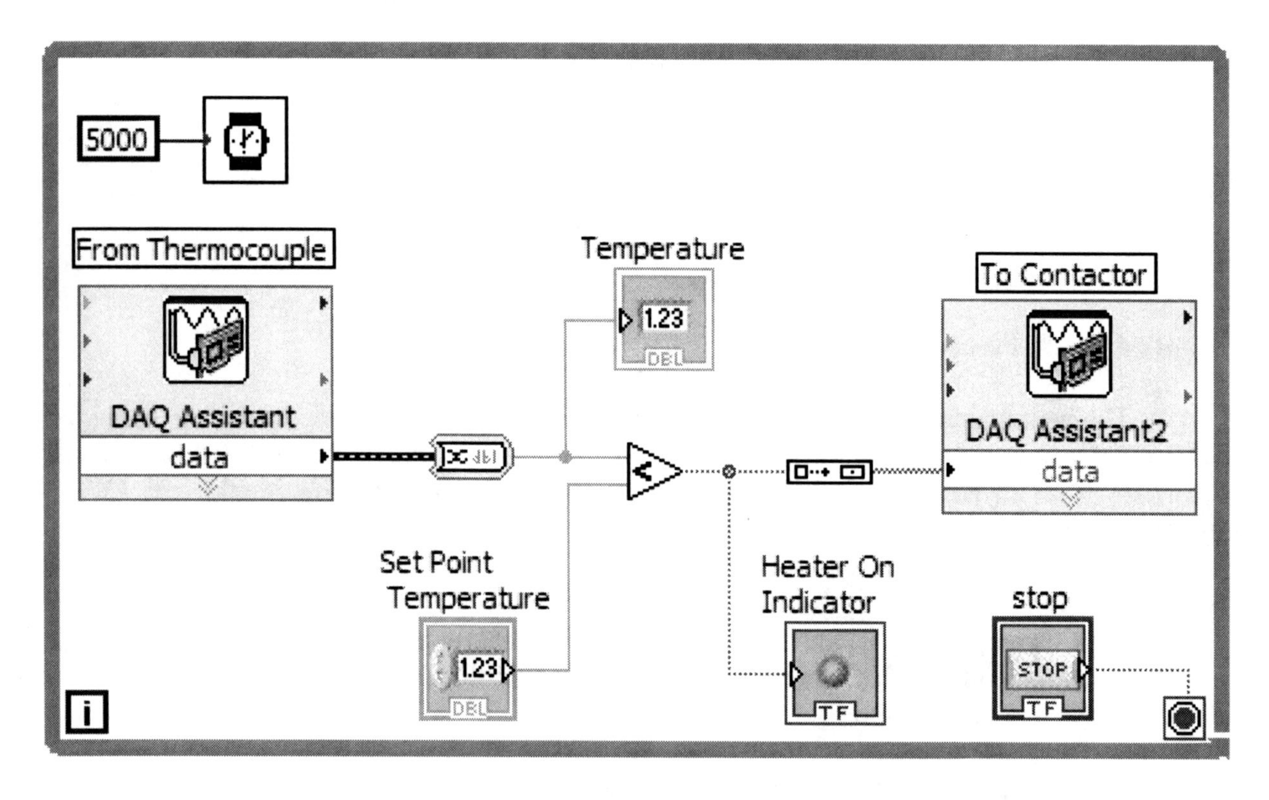

Figure 4-3-3-3 "Block Diagram" VI for comparing the set point °C to the thermocouple output °C and, if warranted, turning on the furnace heater. This figure is in color on the book's front cover.

3) Create the "While Loop" by left-clicking on "View", "Functions Palette", "Express", and "Execution Control". Then left-click, hold, and drag it onto the "Block Diagram". Drag out its sides to make space for inserting blocks.

4) Left-click on “View”, “Functions Palette”, “Programming”, and “Timing”. Then left-click, hold, and drag the “Wait (ms)” function block into the “While Loop”.

5) Set the waiting time for the “Wait (ms)” to 5000 milliseconds by left-clicking “View”, “Express”, and “Arithmetic and Comparison”. Then left-click, hold, and drag the “Numeric Constant” block to the left side of the “Wait (ms)”. If the “Numeric Constant” block does not automatically connect to the “Wait (ms)” run a dataflow wire from it to the “Wait (ms)”. Put 5000 in the “Numeric Constant” block.

6) Create the “DAQ Assistant” function block for inputting thermocouple data by left-clicking on “View”, “Functions Palette”, “Express”, and “Input”. Then left-click, hold, and drag the “DAQ Assistant” function block into the “While Loop”.

7) In the appearing “DAQ Assistant” windows left-click on “Acquire Signals”, “Analog Input”, “Temperature”, “Thermocouple”, “Dev1 (SUB-6210)”, “ai0”, and “Finish”.

8) Leave the default values in the “DAQ Assistant” window. See Figure 4-3-3-4.

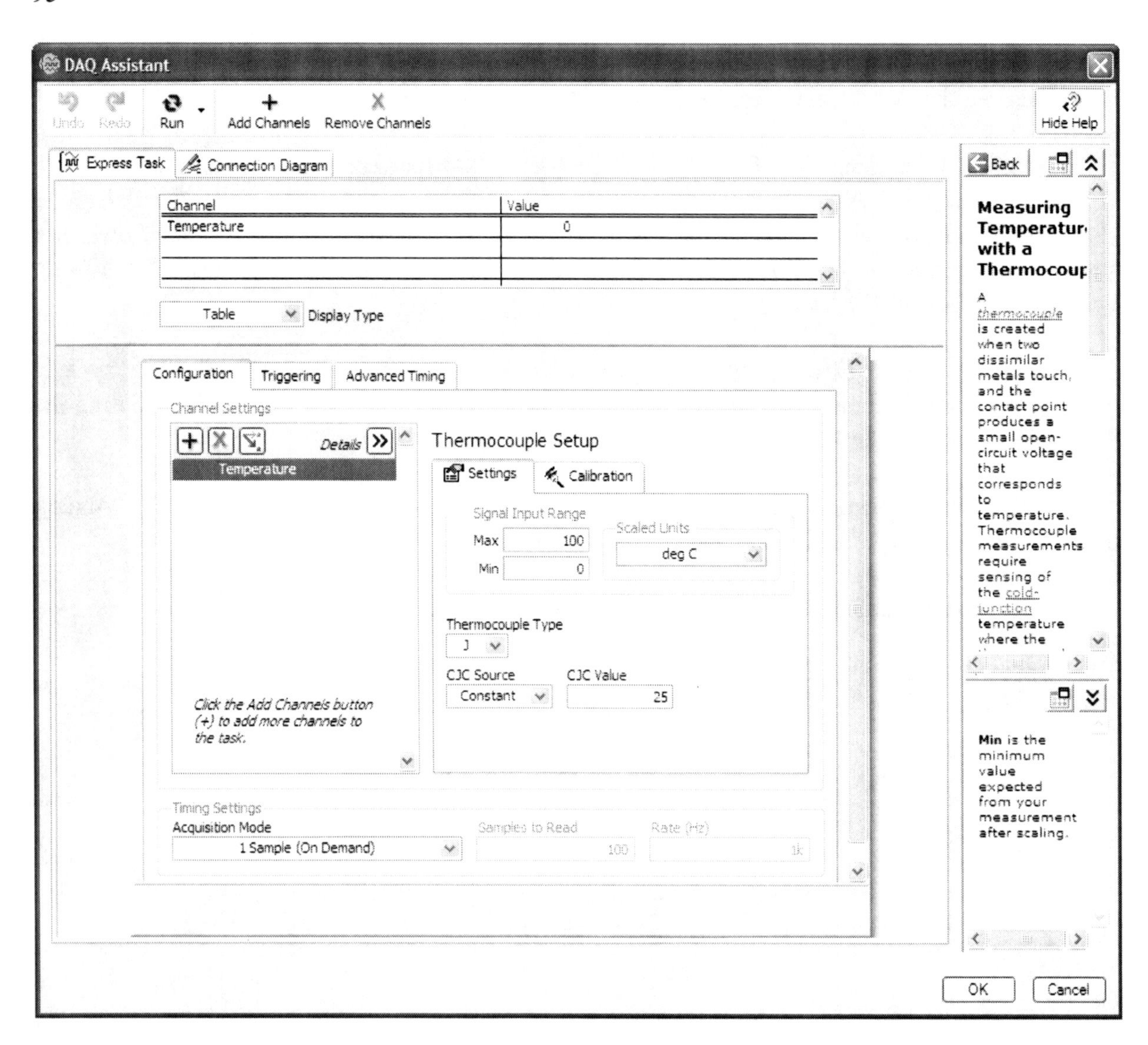

Figure 4-3-3-4 "DAQ Assistant" window for entering thermocouple data.

9) Still in the "DAQ Assistant" window, left-click on "Connection Diagram" to show how the DAQ would be connected. The appearing diagram, Figure 4-3-3-5, should agree with Figure 4-3-3-1.

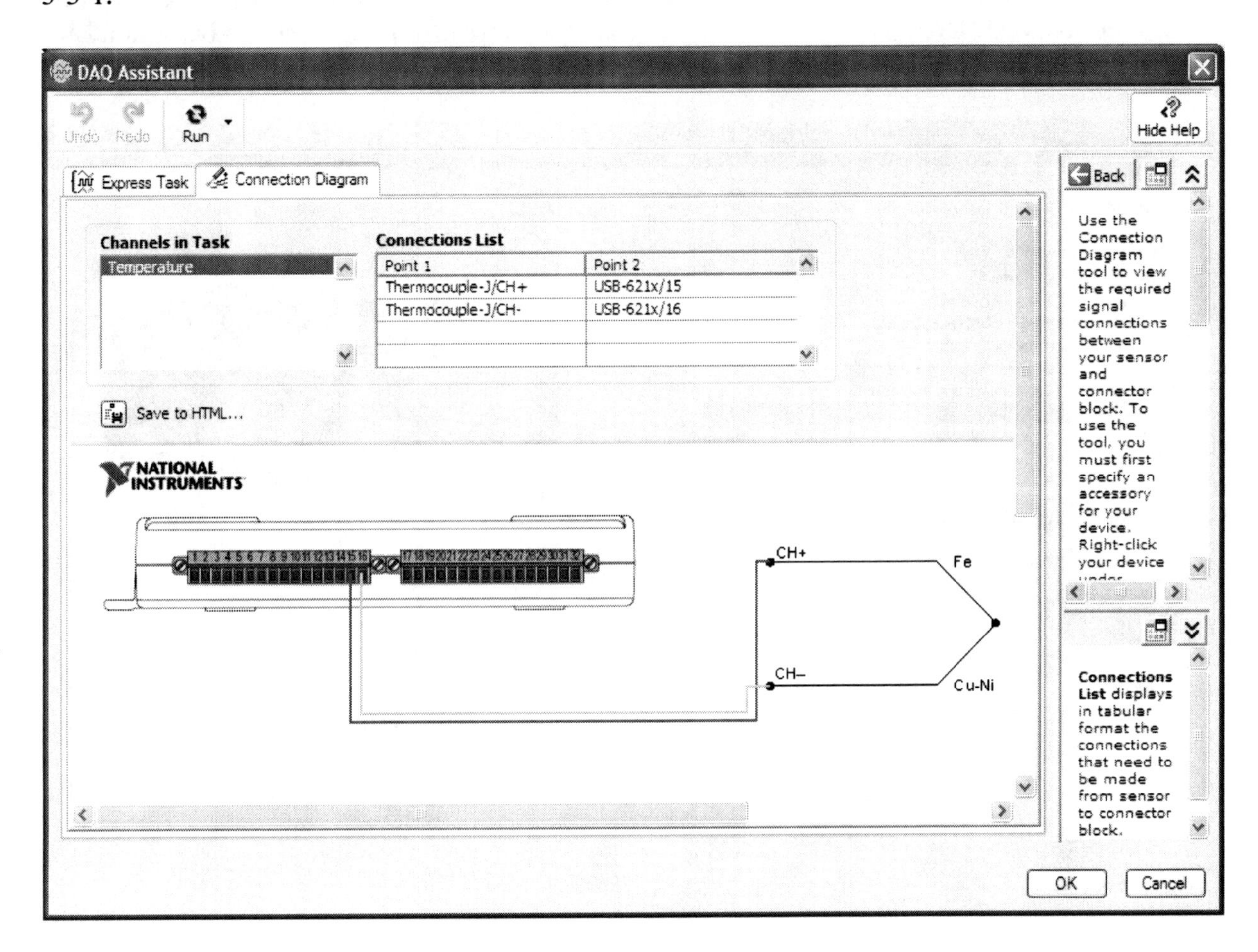

Figure 4-3-3-5 Thermocouple connection diagram.

10) Put the label "From Thermocouple" above the "DAQ Assistant" function block.

11) Create the "Less?" function block by left-clicking on "View", "Functions Palette", "Express", "Arithmetic & Comparison", and "Express Comparison". Then left-click, hold, and drag the "Less?" function block onto the "Block Diagram".

12) Put the "Set Point Temperature" input block onto the "Front Panel" by left-clicking on "View", "Control Panel", "Express", and "Numeric Controls". Then left-click, hold, and drag the "Numeric Control" input block onto the "Front Panel". Change its label from "Numeric" to "Set Point Temperature".

13) On the "Block Diagram" connect a dataflow wire from "Set Point Temperature" to the bottom terminal of the "Less?" function block.

14) On the "Block Diagram" left-click on "View", "Functions Palette", "Express", and "Signal Manipulation". Left-click, hold, and drag the "Convert from Dynamic Data" function block to between the "DAQ Assistant" and "Less?" function blocks. In the appearing window left-click on "Single Scalar" and "OK". To configure the "Convert from Dynamic Data" function block at a later time, right-click on it and left-click on "Properties". See Figure 4-3-3-6.

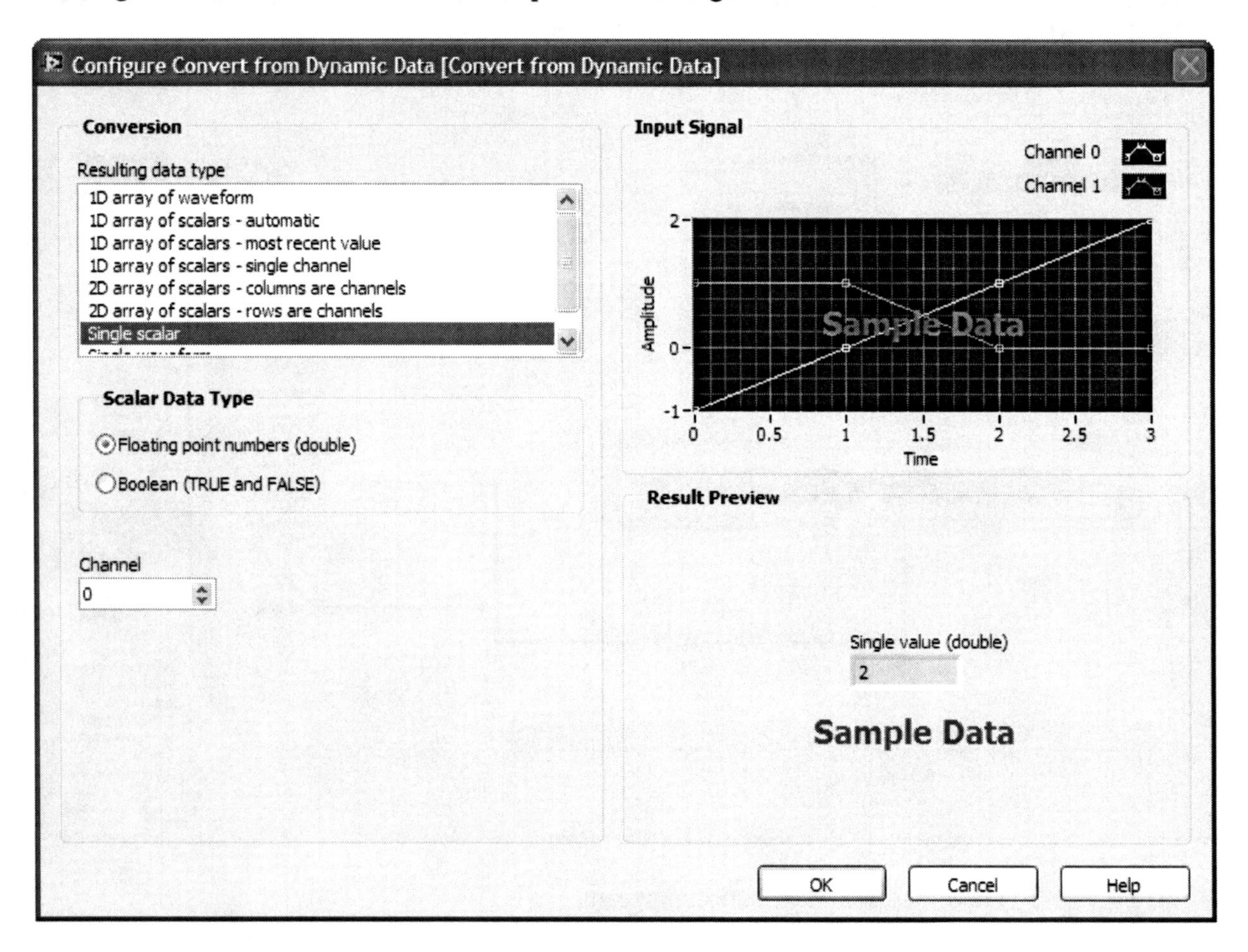

Figure 4-3-3-6 "Convert from Dynamic Data" window.

15) Connect dataflow wires from the "DAQ Assistant" function block to the "Convert Dynamic Data" function block and from the "Convert Dynamic Data" function block to the "Less?" function block.

16) Put the "Temperature" output block onto the "Front Panel" by left-clicking on "View", "Control Panel", "Express", and "Numeric Indicators". Then left-click, hold, and drag the "Numeric Indicator" output block onto the "Front Panel". Change its label from "Numeric" to "Temperature". Connect it by a dataflow wire to the upper "Less?" function block input dataflow wire.

17) Create the "DAQ Assistant" function block for outputting to the amplifier by left-clicking on "View", "Functions Palette", "Express", and "Output". Then left-click, hold, and drag the "DAQ Assistant" function block into the "While Loop".

18) In the appearing "DAQ Assistant" windows left-click on "Generate Signals", "Digital Output", "Line Output", "port1/line0", and "Finish". A "DAQ Assistant" window appears. See Figure 4-3-3-7.

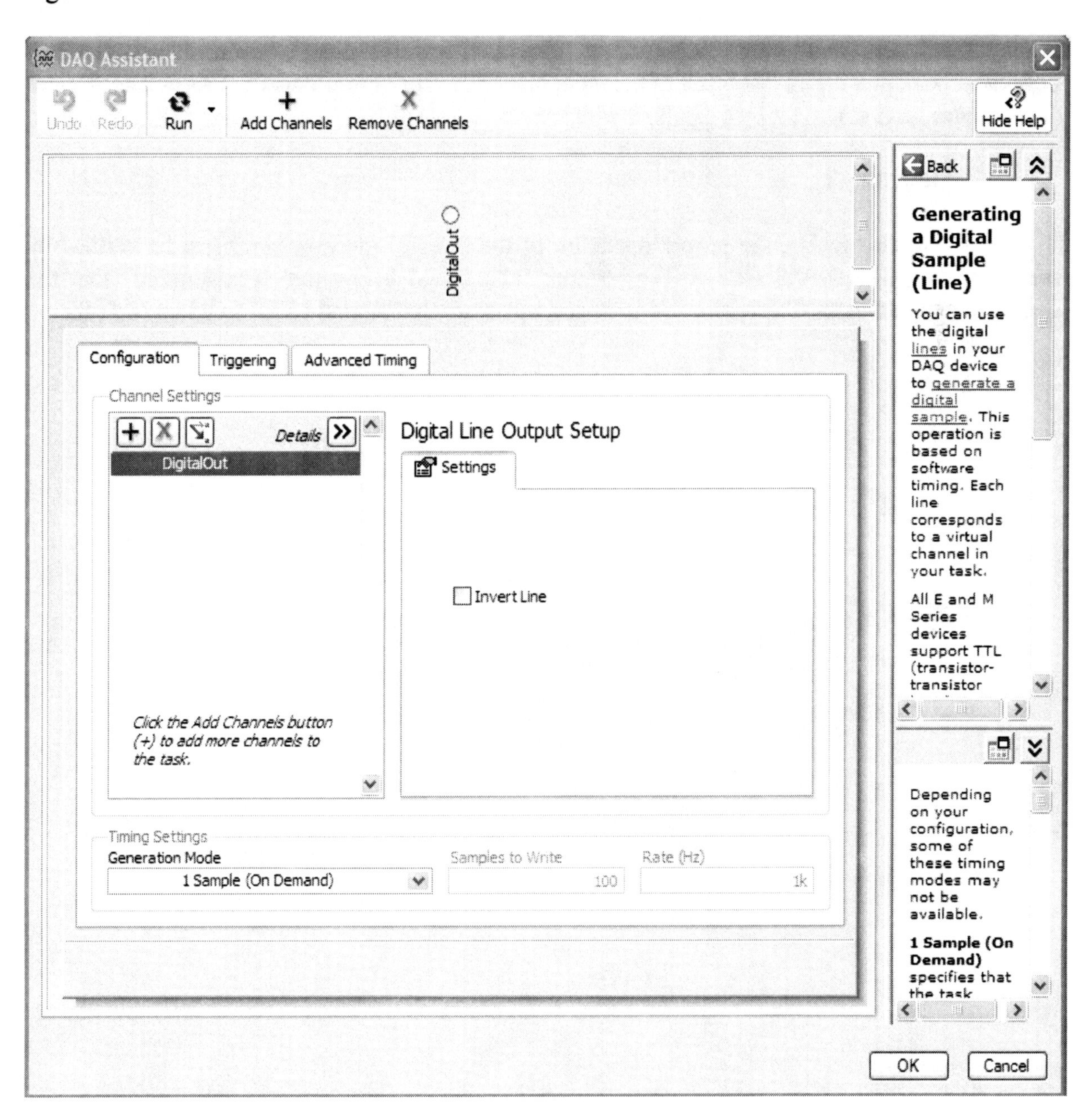

Figure 4-3-3-7 "DAQ Assistant" window for digital outputs.

19 Leave the default values and left-click “OK” on the “DAQ Assistant” window.

20) Label the “DAQ Assistant2” function block as “To Contactor”.

21) On the “Block Diagram” left-click on “View”, “Functions Palette”, “Programming”, and “Array”. Then left-click, hold, and drag the “Build Array” function block to between the “Less?” function block and “DAQ Assistant2” “To Contactor” function block.

22) On the “Front Panel” left-click on “View”, “Controls Palette”, “Express”, and “LEDs”. Left-click, hold, and drag a “Round LED” output block onto the “Front Panel”. Relabel it as “Heater On Indicator” and connect it by a dataflow wire.

23) Run the VI.

24) When running, the proper operation of the “Less?” function block can be verified by raising or lowering the “Set Point Temperature” above and below the “Temperature” that the simulated circuit creates. The raising and lowering will cause the “Round LED” to go on and off.

4.4 CONNECTING LabVIEW TO THE PHYSICAL WORLD THROUGH A NI USB-6009 DAQ

LabVIEW is at its best inputting data it has received from a DAQ (Data Acquisition) device and manipulating that data to form charts and graphs. It can also produce output actions through DAQ devices that have output capabilities.

This section describes the setting up and operating of a NI USB-6009 DAQ device with DAQmx driver software and LabVIEW. The NI USB-6009 is demonstrated in several examples. The purpose of each is to demonstrate the device's capabilities without the distraction of complicated physical setups or LabVIEW VIs. Real applications would probably be more complicated.

Prior to this section, all the examples could be done with only LabVIEW software and a computer. This allowed learning from a demonstration version of LabVIEW. Here the reader without access to a DAQ device will not be able to completely do the examples. However, the reader without access to a DAQ device could setup the examples with a "NI-DAQmx Simulated Device" as was done in Section 4.3 and read through the examples.

The examples presented here could also be done with the less expensive NI USB-6008 rather than the NI USB-6009.

Section 3.3 gives details on purchasing a NI USB-6008 and NI USB-6009.

4.4.1 NI USB-6009 DATA

General	
Product Name	NI USB-6009
Product Family	Multifunction Data Acquisition
Form Factor	USB
Part Number	779026-01
Operating System/Target	Pocket PC, Windows, Linux, Mac 08
DAQ Product Family	B Series
Measurement Type	Voltage
RoHS Compliant	Yes

Analog Input	
Channels	8, 4
Single-Ended Channels	8
Differential Channels	4
Resolution	14 bits
Sample Rate	48 kS/s
Throughput	48 kS/s
Max Voltage	10 V
Maximum Voltage Range	-10 V, 10 V
Maximum Voltage Range Accuracy	138 mV
Minimum Voltage Range	-1 V, 1 V
Minimum Voltage Range Accuracy	37.5 mV
Number of Ranges	8
Simultaneous Sampling	No
On-Board Memory	512 B

Analog Output	
Channels	2
Resolution	12 bits
Max Voltage	5 V
Maximum Voltage Range	0 V, 5 V
Maximum Voltage Range Accuracy	7 mV
Minimum Voltage Range	0 V, 5 V
Minimum Voltage Range Accuracy	7 mV
Update Rate	150 S/s
Current Drive Single	5 mA
Current Drive All	10 mA

Digital I/O	
Bidirectional Channels	12
Input-Only Channels	0
Output-Only Channels	0
Number of Channels	12, 0, 0
Timing	Software
Logic Levels	TTL
Input Current Flow	Sinking, Sourcing
Output Current Flow	Sinking, Sourcing
Programmable Input Filters	No
Supports Programmable Power-Up States?	No
Current Drive Single	8.5 mA
Current Drive All	102 mA
Watchdog Timer	No
Supports Handshaking I/O?	No
Supports Pattern I/O	No
Maximum Input Range	0 V, 5 V
Maximum Output Range	0 V, 5 V

Counter/Timers	
Counters	1
Buffered Operations	No
Debouncing/Glich Removal	No
GPS Synchronization	No
Maximum Range	0 V, 5 V
Max Source Frequency	5 MHz
Minimum Input Pulse Width	100 ns
Pulse Generation	No
Resolution	32 bits
Timebase Stability	50 ppm
Logic Levels	TTL

Physical Specifications	
Length	8.51 cm
Width	8.18 cm
Height	2.31 cm
I/O Connector	Removable Terminal Blocks, Screw-type, Accepts 16 to 28 AWG Wire

Timing/Triggering/Synchronization	
Triggering	Digital
Synchronization Bus (RTSI)	No

Caution: Exceeding maximum voltage or current ratings can damage the NI USB-6009 and/or the computer.

4.4.2 NI USB-6009 PACKAGE INCLUDES:

NI USB-6009 with two removable terminal blocks
USB cable
NI-DAQmx 9.1.7 DVD Driver Software
Booklet 'Getting Started Guide for NI-DAQmx for USB Devices' 12 cm x 12 cm, 8 pages

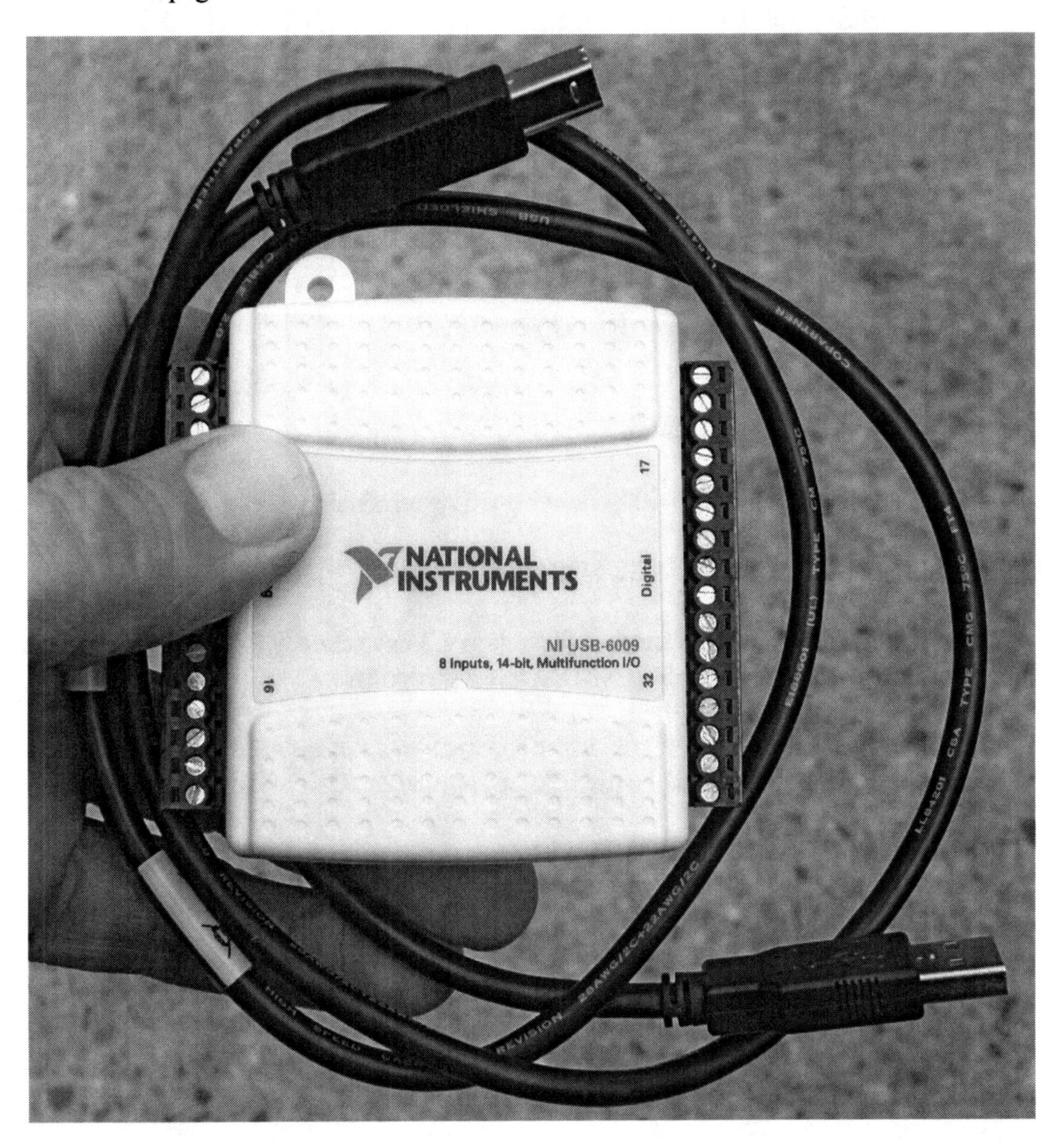

Figure 4-4-2-1 Photo of a NI USB-6009, installed removable terminal blocks, and USB cable. Actual size.

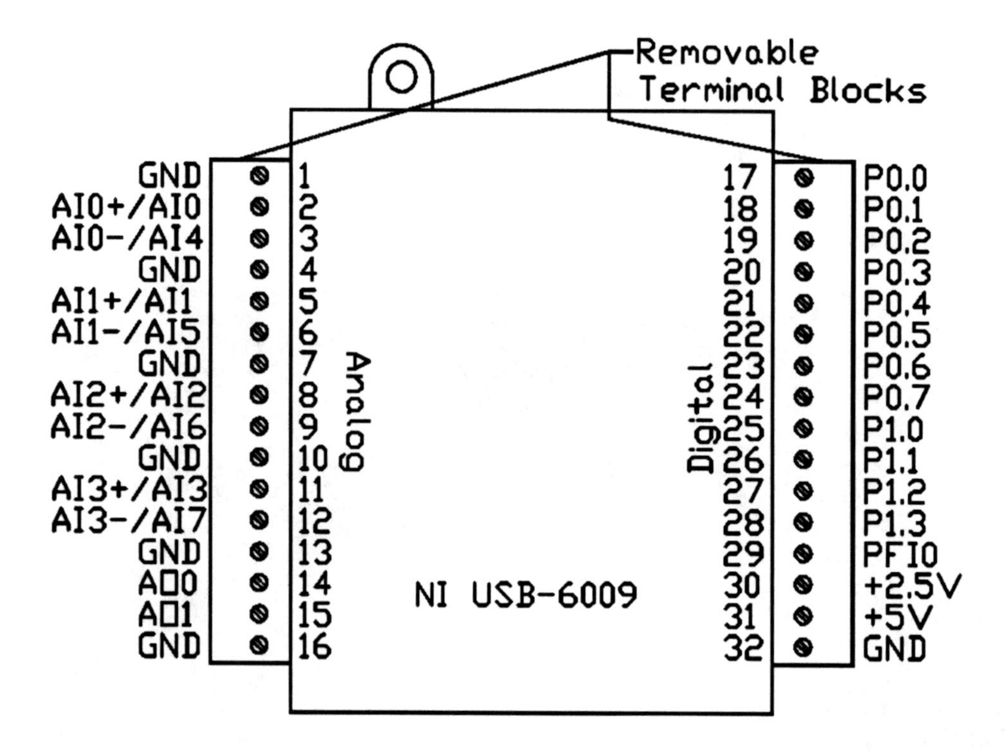

Figure 4-4-2-2 Top view of a NI USB-6009 showing wiring terminals.

4.4.3 INSTALLING NI-DAQmx DRIVER SOFTWARE

Those that did the simulated exercises in Section 4.3 have already followed the procedure of Section 1.6 for downloading and installing NI-DAQmx driver software.

If a correct version of NI-DAQmx driver software was not installed earlier, then the DVD supplied with the NI USB-6009 should be used. After inserting the DVD follow the prompts.

4.4.4 FIRST CONNECTING TO A COMPUTER

1) Start the computer

2) Using the USB cable, connect the NI USB-6009 to the computer. Wiring should not be connected to the NI USB-6009 terminal blocks at this time.

3) The green LED on the NI USB-6009 will flash on and off.

4) Start the "Measurement & Automation" software, the same software that is used to start LabVIEW.

5) On the “Configuration” window left-click on “Devices and Interfaces”. “NI USB-6009”Dev2”” and a green icon should automatically appear under “NI USB-6210 “Dev 1””. LabVIEW will have sensed the new device and put it on the “Devices and Interfaces” list automatically. The NI USB-6210 is left over from the simulated device examples of Section 4.3. Notice that the “NI USB-6009 “Dev2”” icon is green, indicating it is an actual device, while the NI USB-6210 is yellow, indicating it is a simulated device. This can be seen in Figure 4-4-4-1.

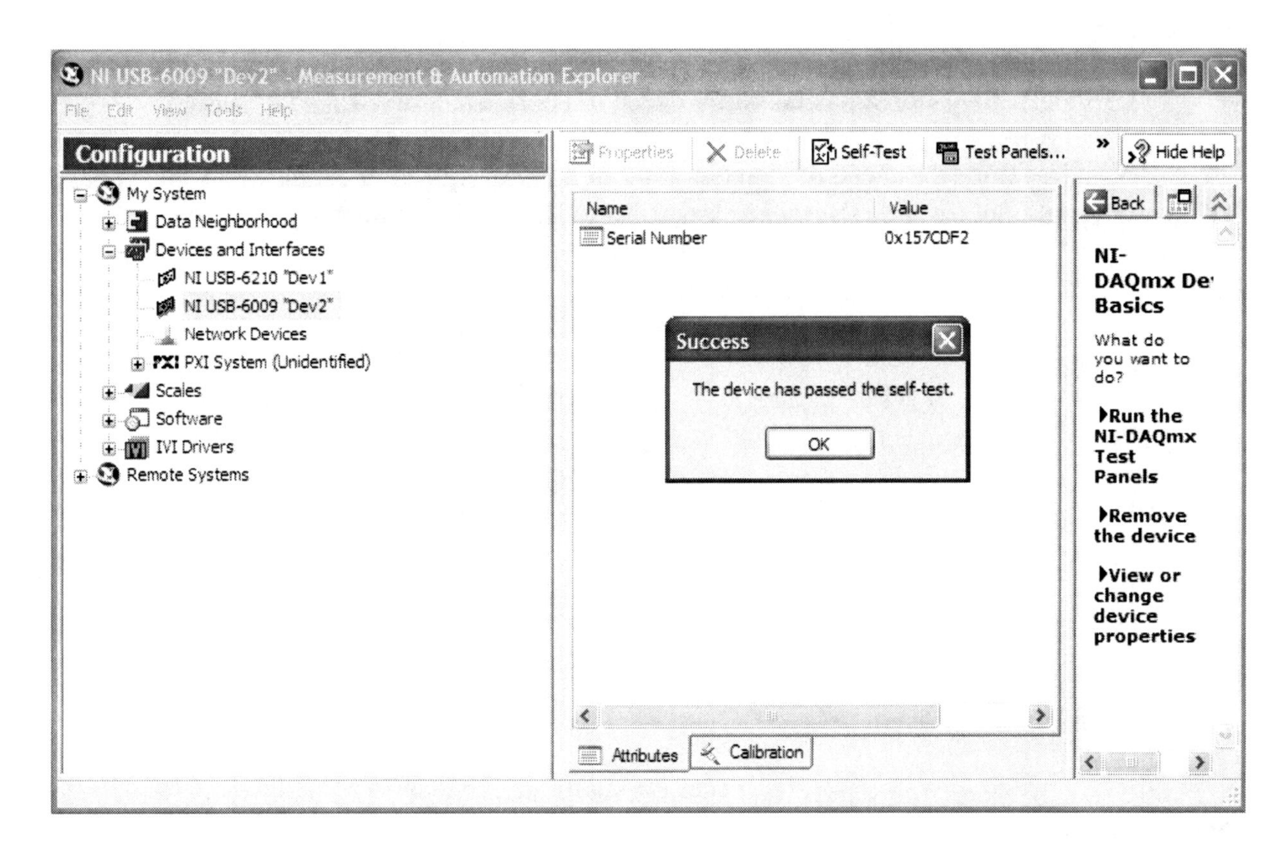

Figure 4-4-4-1 “Devices and Interfaces” list and “Success” window.

6) Confirm that LabVIEW has accepted the “NI USB-6009” by right-clicking on “NI USB-6009 “Dev2”” and left-clicking “Self-Test”. A small “Success” window should appear that states “The device has passed the self-test.” See Figure 4-4-4-1.

7) The proper functioning of the "NI USB-6009" and its inputs can also be checked with the "Test Panels" feature. This can be accessed by right-clicking on "NI USB-6009" and left-clicking on "Test Panels". It can check analog inputs & outputs, digital inputs & outputs and the counter.

8) For an example of "Test Panels" use, check the AI0 (analog input at screw terminal 2) relative to GND (screw terminal 1) for the voltage from a low voltage DC source:

Equipment:
NI USB-6009
1.5 VDC battery/switch/potentiometer circuit or equivalent low voltage DC supply

a) Build the circuit of Figure 4-4-4-2 or build an equivalent circuit with a low voltage DC power supply. Set the potentiometer or low voltage DC supply to output about .6 VDC.

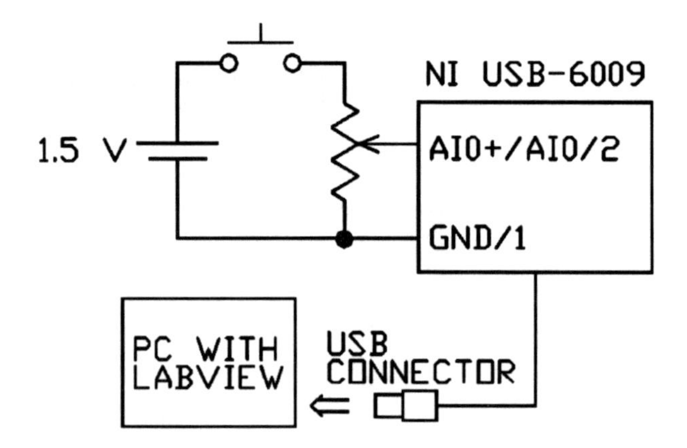

Figure 4-4-4-2 Switchable low voltage DC supply circuit connected to the NI USB-6009.

b) Start "Test Panels" and left-click on "Analog Input". The window can be seen in Figure 4-4-4-3.

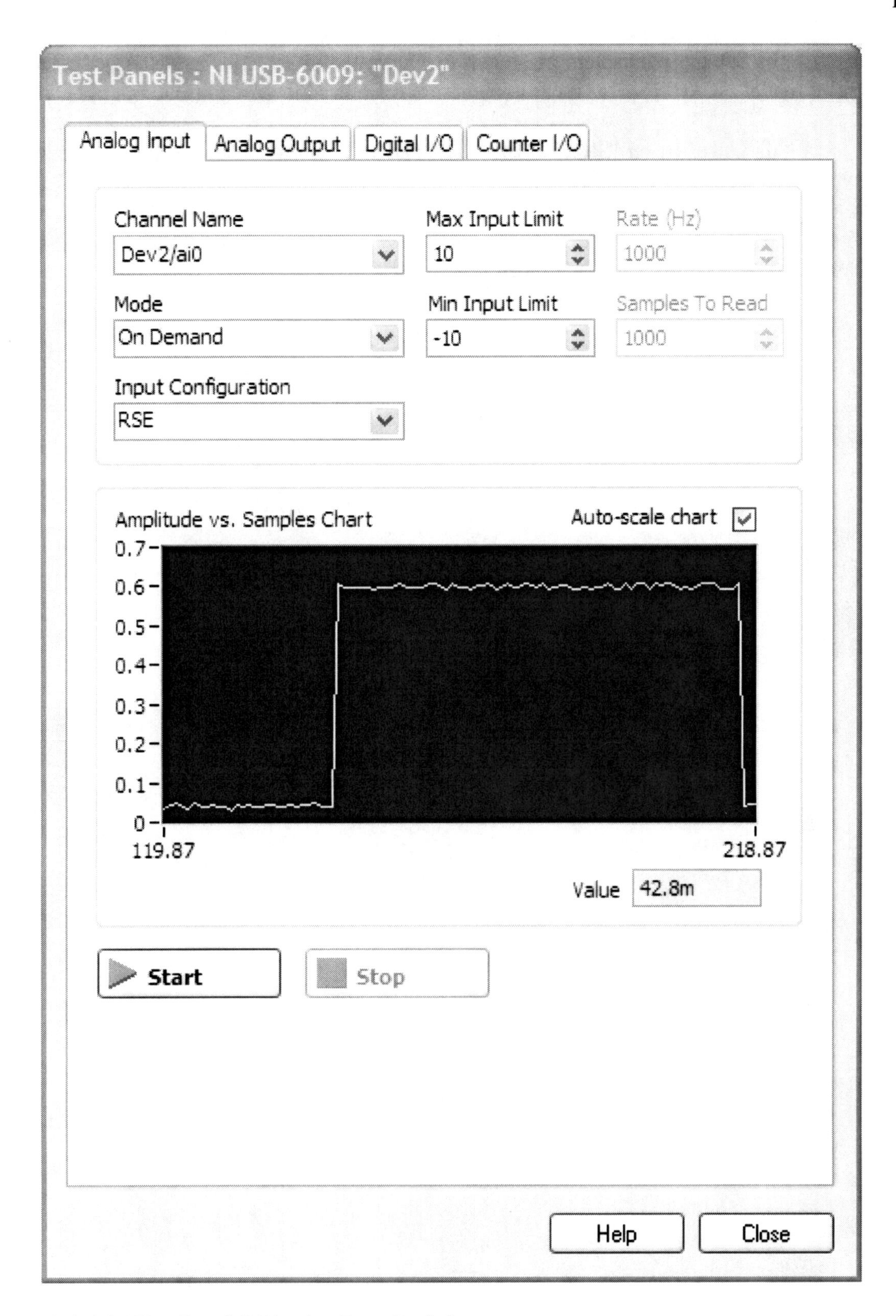

Figure 4-4-4-3 “Test Panels” “Analog Input” window.

c) Set the parameters as shown in Figure 4-4-4-3. “RSE” stands for “referenced single-ended mode”. With “RSE” all measurements are referenced to ground (GND).

d) Left-click on “Start” in the “Test Panels” window and close the push button on the potentiometer circuit. A trace like that in Figure 4-4-4-3 should appear. Since the “Test Panels” window was set on “Auto-scale chart”, after a short period the chart will show only the voltage range that it is receiving. A .6 VDC from the battery/switch/potentiometer circuit will appear as a jagged waveform varying from .55 to .75 VDC.

4.4.5 ANALOG INPUT AND OUTPUT

The NI USB-6009 has eight analog input lines and two analog output lines.

Analog inputs can be differential or relative to ground ("RSE" or referenced single-ended). There are a maximum of four differential inputs and a maximum of eight relative to ground inputs. Input voltages to each line can be a maximum of +10 V and a minimum of -10 V.

The analog inputs can be triggered on with the PFI0 line input. With a rising or falling PFI0 input voltage, the analog input will begin acquiring data. The trigger is configured when setting up the "DAQ Assistant" analog input.

Analog outputs are relative to ground. They can produce from 0 to +5 V.

4.4.5.1 Measurement and Display of an Analog Voltage

Problem:

Use LabVIEW and the NI USB-6009 to measure the real-time voltage of a varying source and of an open terminal.

Equipment:

Same as in Section 4.4.4 page 107

NI USB-6009's inputting a referenced to ground analog voltage into LabVIEW is demonstrated here.

Solution:

1) Build the circuit of Figure 4-4-5-1-1.

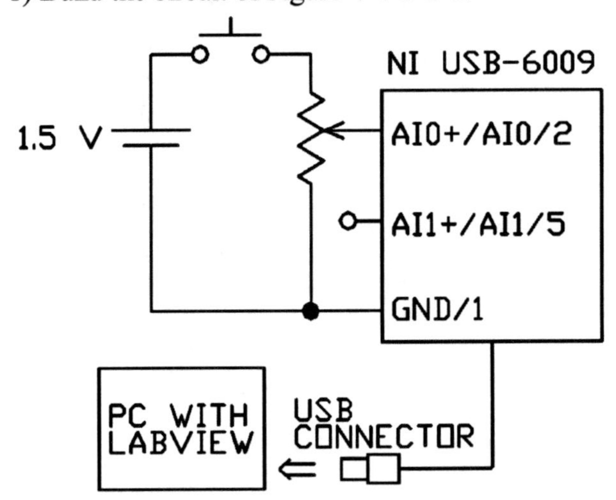

Figure 4-4-5-1-1 Switchable low voltage DC supply circuit and open terminal connected to the NI USB-6009.

2) Create the VI of the "Front Panel" of Figure 4-4-5-1-2 and the "Block Diagram" of Figure 4-4-5-1-3.

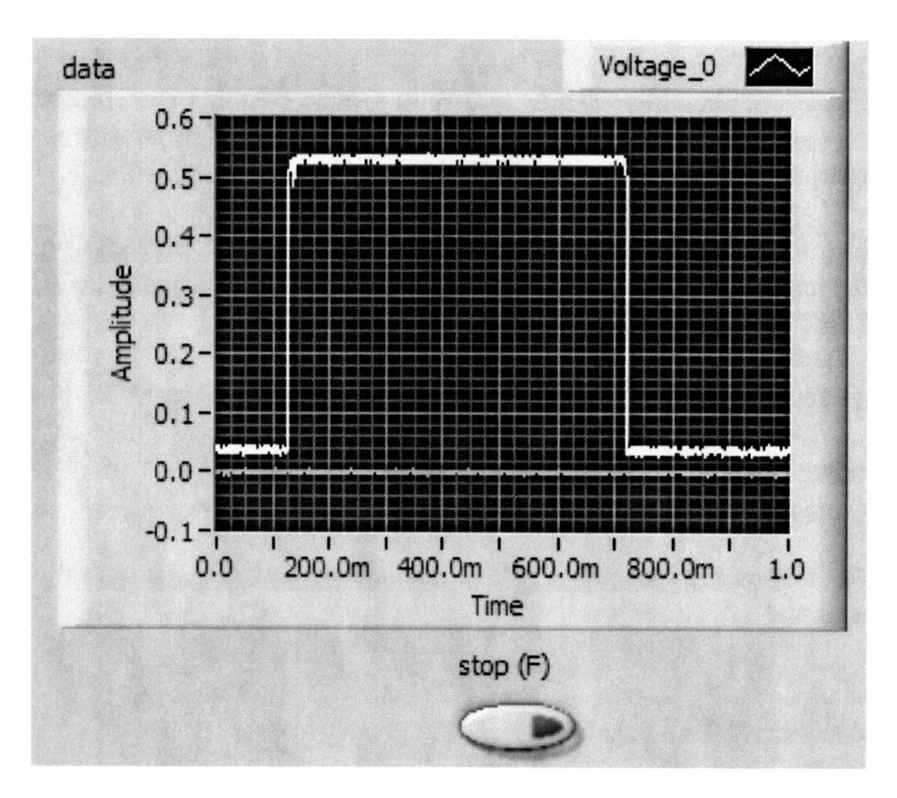

Figure 4-4-5-1-2 "Front Panel" for reading an analog input from the NI USB-6009. The push button was pushed at .12 seconds and released at .72 seconds.

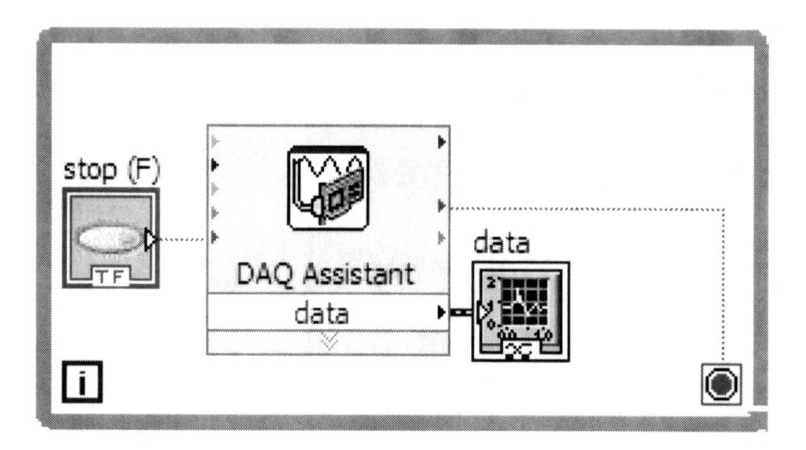

Figure 4-4-5-1-3 "Block Diagram" for reading an analog input from the NI USB-6009.

3) To create the “DAQ Assistant“ function block left-click on “View”, “Functions Palette”, “Express”, and “Input”. Then left-click, hold, and drag it onto the “Block Diagram”.

4) In the appearing window left-click on “Acquire Signals”, “Analog Input”, “Voltage”, “Dev2 (SUB-6009)”, “ai0”, “Ctrl” “ai1”, and “Finish”.

5) In the appearing “DAQ Assistant” window left-click on “RSE” under “Terminal Configuration”, and “Continuous Samples” under “Acquisition Mode”. See Figure 4-4-5-1-4.

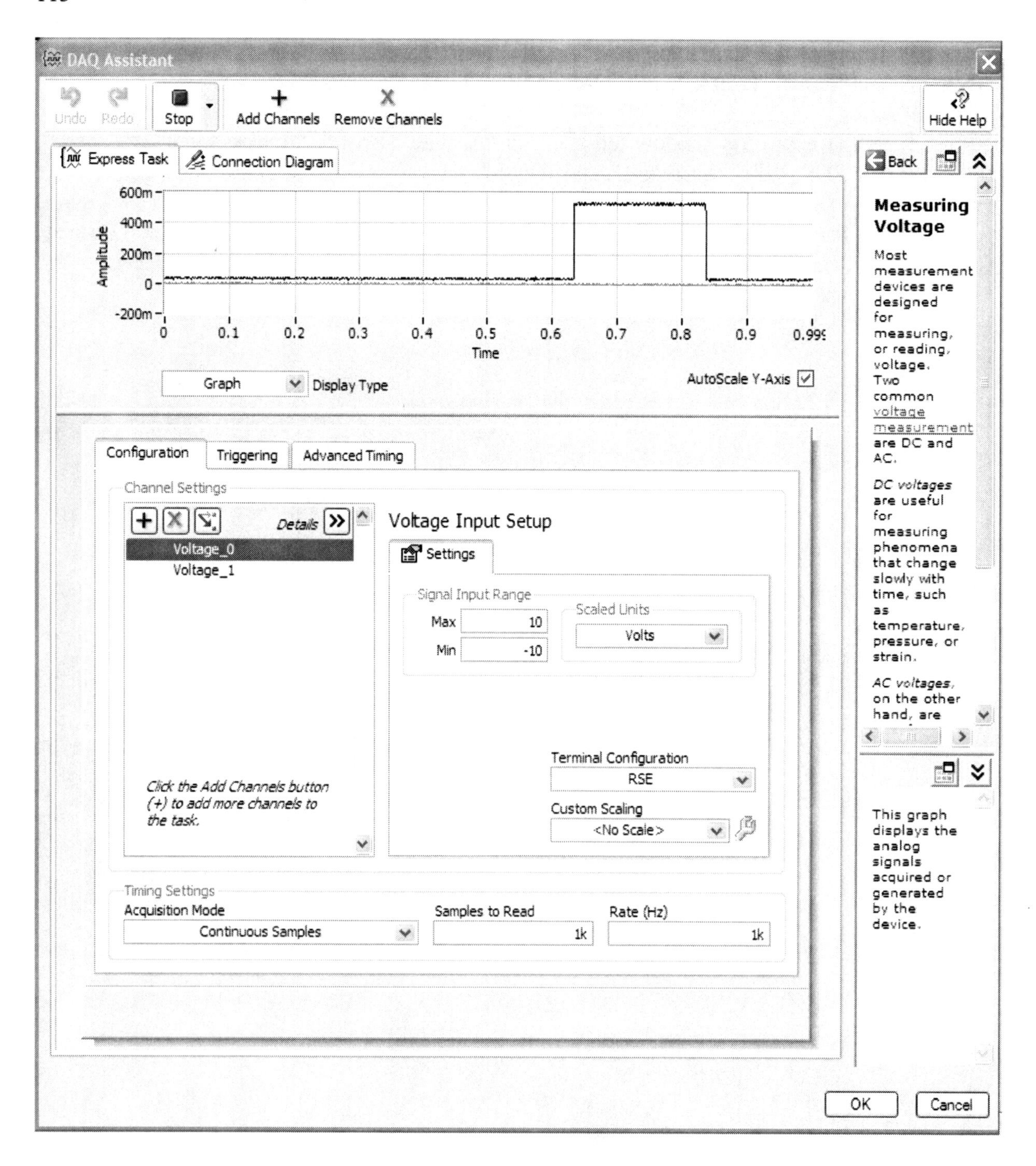

Figure 4-4-5-1-4 “DAQ Assistant” window for analog voltage inputs into the NI USB-6009.

6) Still in the "DAQ Assistant" window, left-click on "Connection Diagram" to show how the DAQ "Voltage_0"/"ai0" should be connected. The appearing diagram should agree with Figure 4-4-4-2.

7) Still in the "DAQ Assistant" window, left-click on "Express Task" and "Run" to test the circuits and NI USB-6009. Two test traces appear, the "ai0" input and the "ai1" input. The "ai0" input has been made to go up and down by pressing and releasing the push button. The "ai1" is the voltage that the unconnected "ai1" sees relative to "GND". The voltage seen here at the unconnected input is 0 VDC, but it does not have to be 0 VDC. Commonly a small voltage will appear across an unconnected input. See Figure 4-4-5-1-4.

8) Left-click on "Stop" and the top of the "DAQ Assistant" window and "OK" at the bottom of the "DAQ Assistant" window.

9) A window appears that offers to place the "DAQ Assistant" function block in a "While Loop". Left-click on "Yes". This will make the VI measure voltage continuously.

10) Right-click on "data" in the "DAQ Assistant" function block, then left-click on "Create" and "Graph Indicator". This places the data graph seen in Figure 4-4-5-1-2.

11) Run the VI.

12) Press and release the push button to make the waveform seen in Figure 4-4-5-1-2. Note the 0 VDC trace. As in the "DAQ Assistant" test window, a trace indicating 0 VDC or some other value may appear for an unconnected analog input.

4.4.5.2 Creation of an Analog Output Voltage

Problem:

Use LabVIEW and the NI USB-6009 to create a 0 to 5 VDC voltage that is controlled by a "Front Panel" rotary knob.

Equipment:

NI USB-6009

VOM set to read DC voltage

LabVIEW's directing a NI USB-6009 to produce a controllable analog DC output voltage is demonstrated here.

Solution:

1) Build the circuit of Figure 4-4-5-2-1.

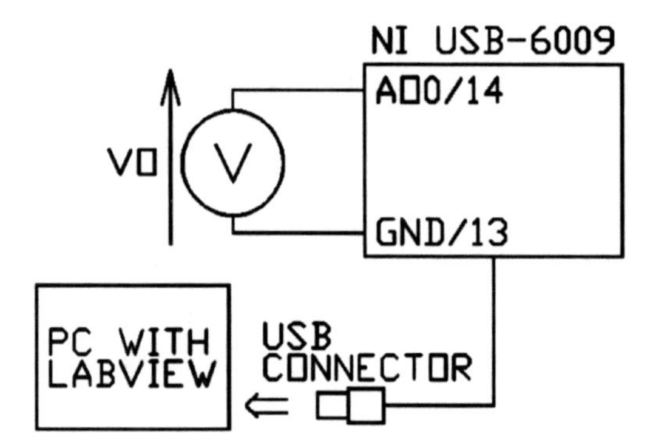

Figure 4-4-5-2-1 Measuring analog output voltage from the NI USB-6009.

2) Create the VI of the "Front Panel" of Figure 4-4-5-2-2 and the "Block Diagram" of Figure 4-4-5-2-3.

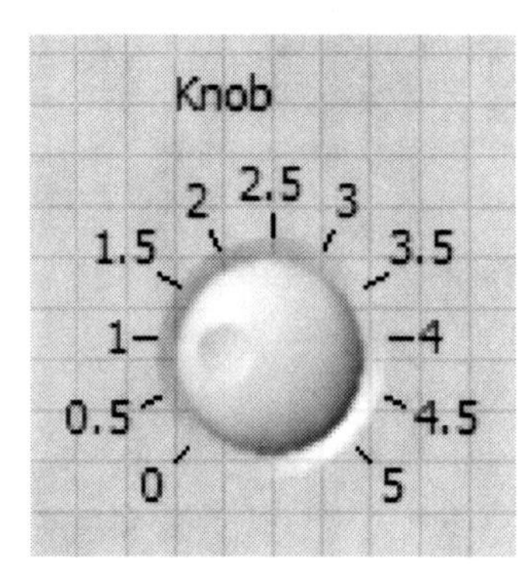

Figure 4-4-5-2-2 "Front Panel" rotary "Knob" input block for controlling analog NI USB-6009 output voltage.

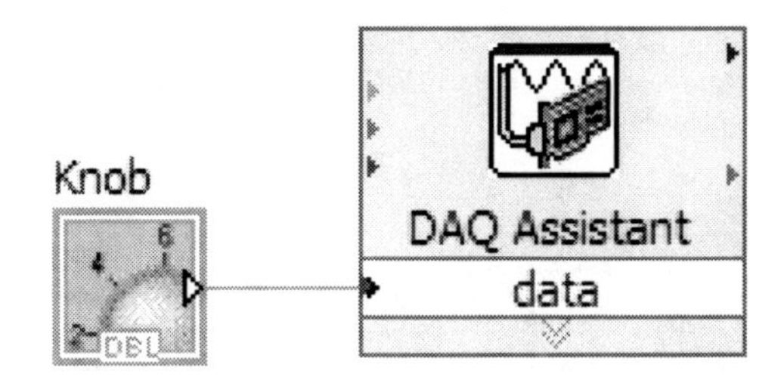

Figure 4-4-5-2-3 “Block Diagram” for producing output voltage.

3) To create the “DAQ Assistant“ function block, left-click on “View”, “Functions Palette”, and “Output”. Then left-click, hold, and drag it onto the “Block Diagram”.

4) In the appearing window left-click on “Generate Signals”, “Analog Output”, “Voltage”, “Dev2 (USB-6009)”, “ao0”, and “Finish”.

5) In the appearing “DAQ Assistant” window leave the default values. See Figure 4-4-5-2-4.

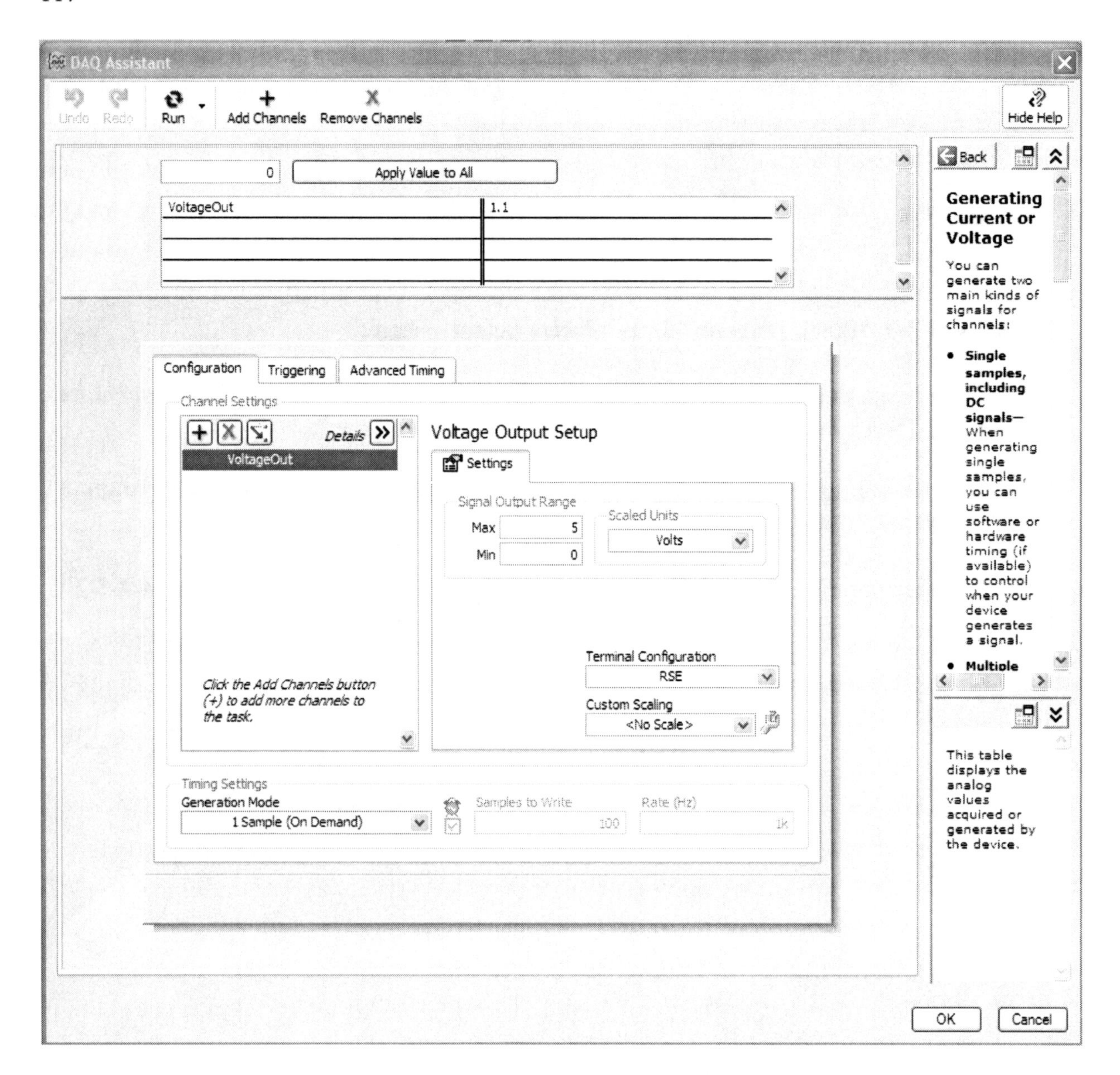

Figure 4-4-5-2-4 "DAQ Assistant" window for analog output.

6) Still in the "DAQ Assistant" window, left-click on "Run". This should produce a voltage of 1.1 volts across the VOM. This "Run" is for testing the NI USB-6009 and the analog output. Later, when the VI is run, the voltage will be controlled from the "Front Panel" rotary "Knob" input block as shown in Figure 4-4-5-2-2.

7) Left-click "Stop" at the top of the "DAQ Assistant" window and "OK" at the bottom of the "DAQ Assistant" window.

8) On the “Front Panel” left-click on “View”, “Controls Palette”, “Express”, “Numeric Controls”. Then left-click and drag a “Knob” input block onto the “Front Panel”.

9) Right-click on the “Front Panel” “Knob” input block and left-click on “Properties” and “Scale”. Set the “Scale Range” to 0 “Minimum” and 5 “Maximum”. Then left-click on “OK”. See Figure 4-4-5-2-5.

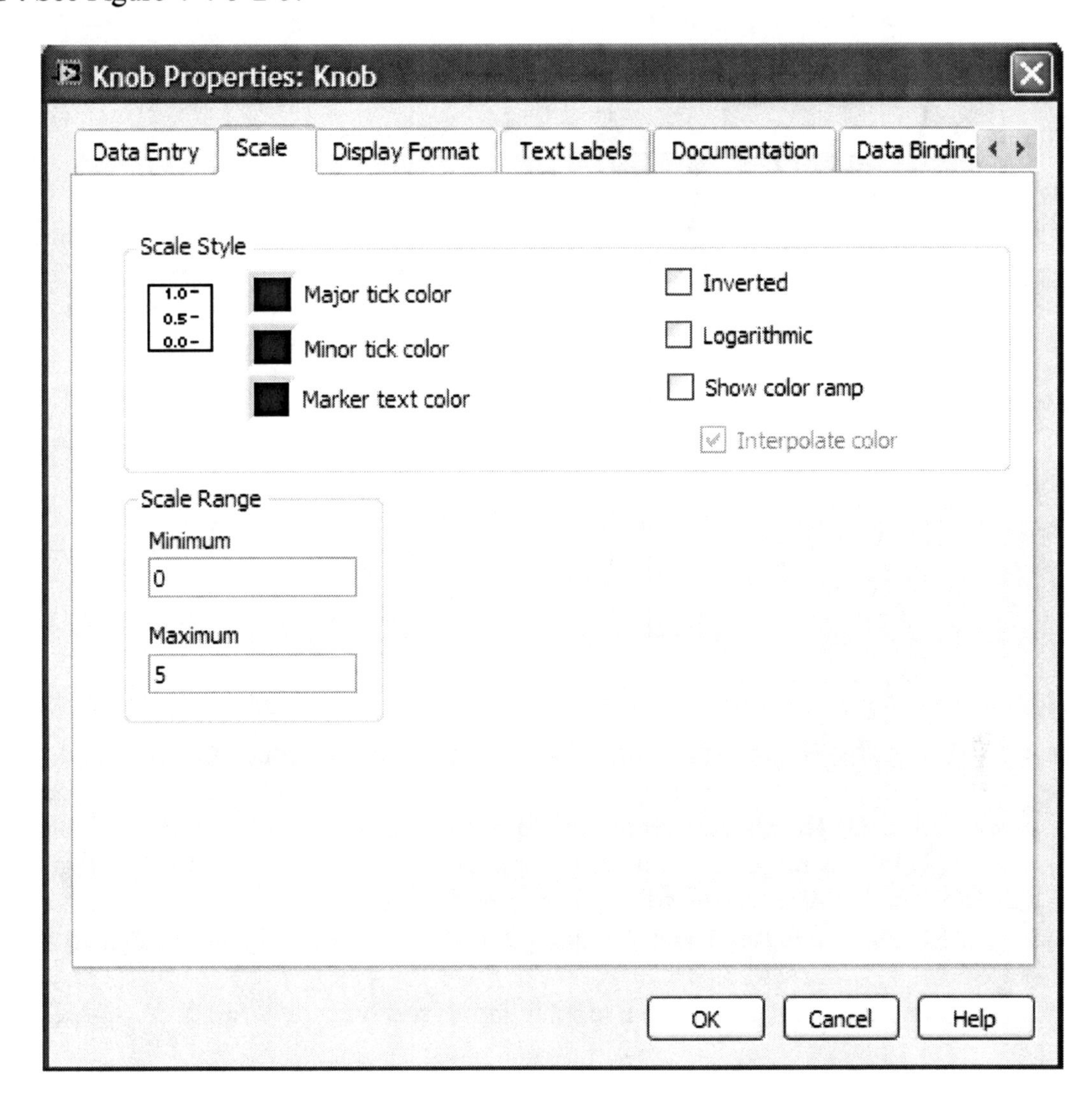

Figure 4-4-5-2-5 Rotary “Knob Properties” window, “Scale” tab.

10) On the “Block Diagram” connect the “Knob” input block to the “DAQ Assistant” function block “data”.

11) Left-click on the “Run Continuously” command (icon to the right of the “Run” command). This continuously restarts the VI so that the “Knob” input block will change the NI USB-6009 output voltage whenever it is turned.

4.4.6 DIGITAL INPUT AND OUTPUT

The NI USB-6009 has 12 digital input/output lines, P0.0, P0.1, P0.2, P0.3, P0.4, P0.5, P0.6, P0.7, P1.0, P1.1, P1.2, and P1.3. Each of these can be configured through "DAQ Assistant" software as either an input or output. Figure 4-4-6-1 shows some possible digital input and output connections.

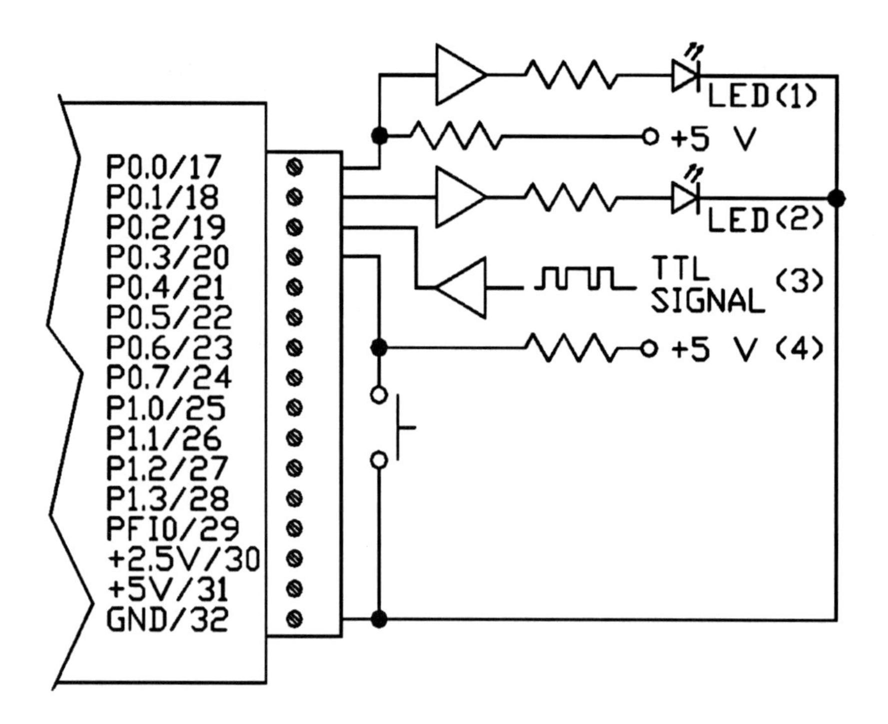

Figure 4-4-6-1 Possible NI USB-6009 digital input and output connections:

(1) P0.0 configured as an open collector digital output sink to an Op Amp that powers an LED. The Op Amp is needed to power a medium to large LED.

(2) P0.1 configured as an active digital output source to an Op Amp that powers an LED. The Op Amp is needed to power a medium to large LED.

(3) P0.2 configured as a digital input receiving a TTL signal through an isolation amplifier.

(4) P0.3 configured as a digital input receiving a 0 or 5 V signal from a pushbutton.

To protect the NI USB-6009 and other DAQs:

1) If a digital line is configured in LabVIEW as an output, do not connect it to an external source.

2) If a digital line is configured in LabVIEW as an output, do not exceed its current limit. See data in Section 4.4.1.

3) If a digital line is configured in LabVIEW as an input do not exceed its voltage limit. Note the digital voltage limit is less than the analog voltage limit. See data in Section 4.4.1.

4.4.6.1 Reading Digital Inputs

Caution: Do not run a VI with the circuit connected before the digital inputs are properly configured in LabVIEW.

Problem:

Use LabVIEW and the NI USB-6009 to read pushbutton positions.

Equipment:

NI USB-6009
Three pushbuttons
Three 4.7 kΩ resistors

NI USB-6009's inputting a digital signal into LabVIEW and LabVIEW's conversion of a 1D Boolean array to individual Boolean scalars are demonstrated here.

Solution:

1) Build the circuit of Figure 4-4-6-1-1.

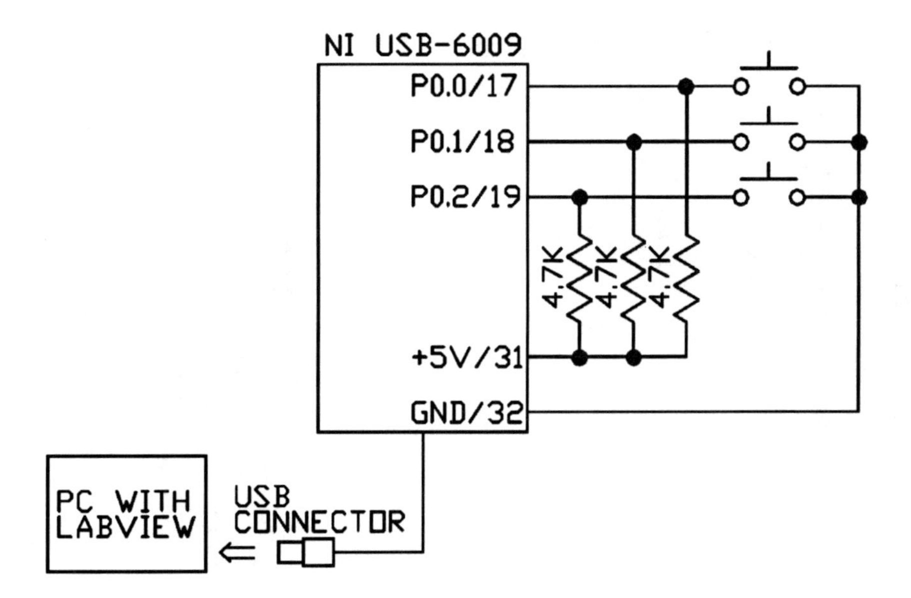

Figure 4-4-6-1-1 Three on/off pushbuttons inputting data to the NI USB-6009.

2) Create the VI of the "Front Panel" of Figure 4-4-6-1-2 and the "Block Diagram" of Figure 4-4-6-1-3.

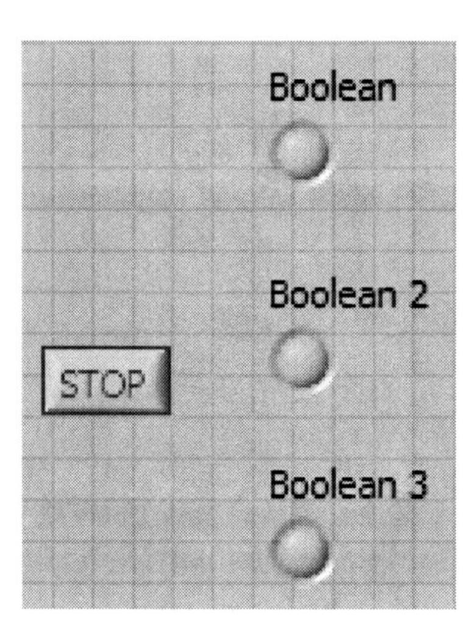

Figure 4-4-6-1-2 "Front Panel" with indicating LEDs.

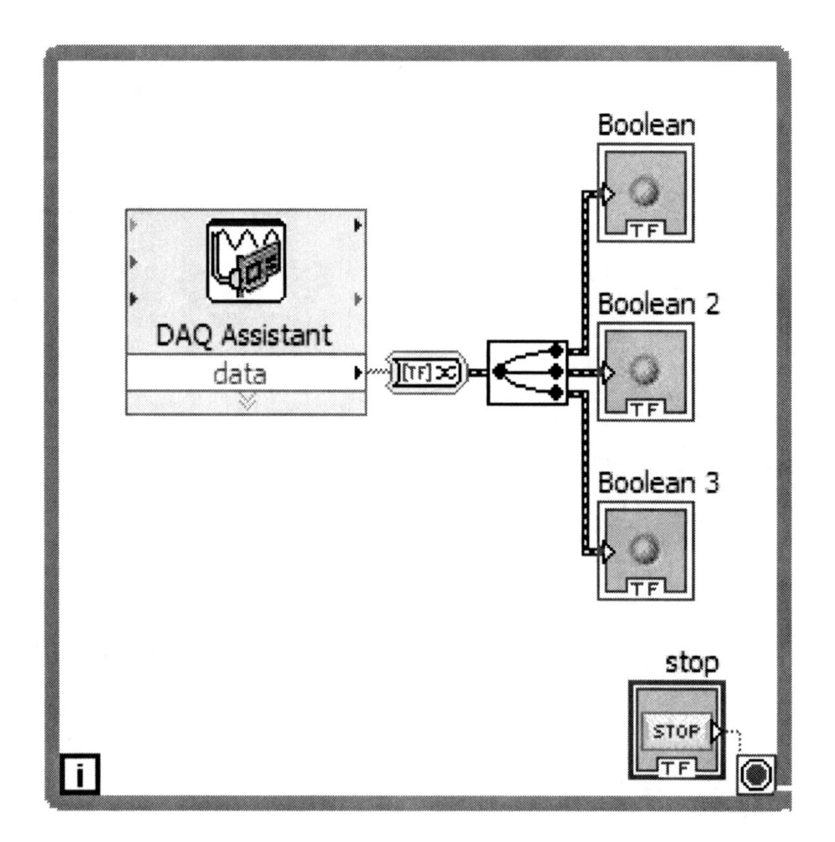

Figure 4-4-6-1-3 "Block Diagram" for receiving pushbutton positions.

3) To create the “DAQ Assistant“ function block left-click on “View”, “Functions Palette”, “Express”, and “Input”. Then left-click, hold, and drag it onto the “Block Diagram”.

4) In the appearing window left-click on “Acquire Signals”, “Digital Input”, “Line Input”, “Dev2 (SUB-6009)”, “port0/line0”, “Ctrl” “port0/line1”, “Ctrl” “port0/line2”, and “Finish”.

5) In the appearing “DAQ Assistant” window leave the default values. See Figure 4-4-6-1-4.

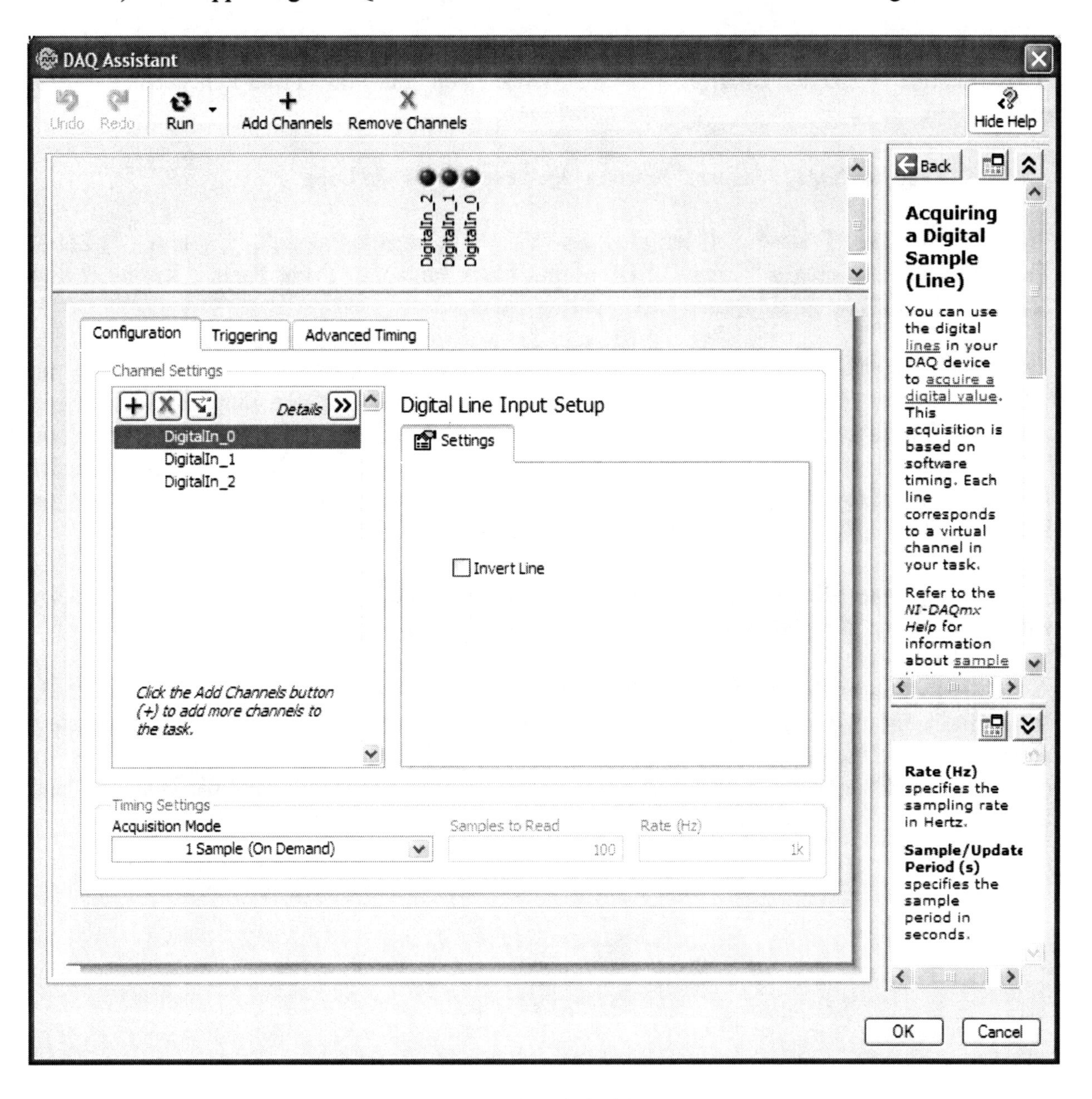

Figure 4-4-6-1-4 “DAQ Assistant” window for digital inputs.

6) Still in the "DAQ Assistant" window, left-click on "Run". This turns on the green LEDs above "DigitalOut_0", "DigitalOut_1", and "DigitalOut_2". These can be turned off by pressing closed their respective pushbuttons. This "Run" is for testing the NI USB-6009 and the digital input circuitry. When the VI is run, "Front Panel" "Round LED" indicators will be turned on and off. See Figure 4-4-6-1-2.

7) Left-click "OK" at the bottom of the "DAQ Assistant" window.

8) Put a "While Loop" on the "Front Panel" by left-clicking on "View", "Functions Palette", "Express", and "Execution Control". Drag the "While Loop" onto the "Front Panel" and stretch it out.

9) Drag the "DAQ Assistant" function block into the "While Loop".

10) On the "Front Panel" left-click on "View", "Controls Palette", "Express", "LEDs". Then left-click and drag a "Round LED" output block onto the "Front Panel". Repeat this to drag on a second and third "Round LED" output block.

11) The "Round LED" output blocks receive simple scalar Boolean values, but the "DAQ Assistant" function block, as configured here, produces a one dimensional Boolean array. A "Split Signals" function block is used to make the conversion. Get it by left-clicking on "View", "Functions Palette", "Express", and "Signal Manipulation". Drag the "Split Signals" function block into the "While Loop". Add two more inputs to the "Split Signals" function block by left-clicking on its bottom border and dragging its border down.

12) Connect the Boolean "LED" output blocks to the "Split Signals" function block by dataflow wires as in Figure 4-4-6-1-3.

13) Connect the "DAQ Assistant" function block to the "Split Signals" function block. When the connection is made a "Convert to Dynamic Data" function block automatically appears. Right-click on "Convert to Dynamic Data" and left-click on "Properties" to show its configuration. Change the "Input data type" to "1D array of scalars – multiple channels" and left-click "OK". See Figure 4-4-6-1-5.

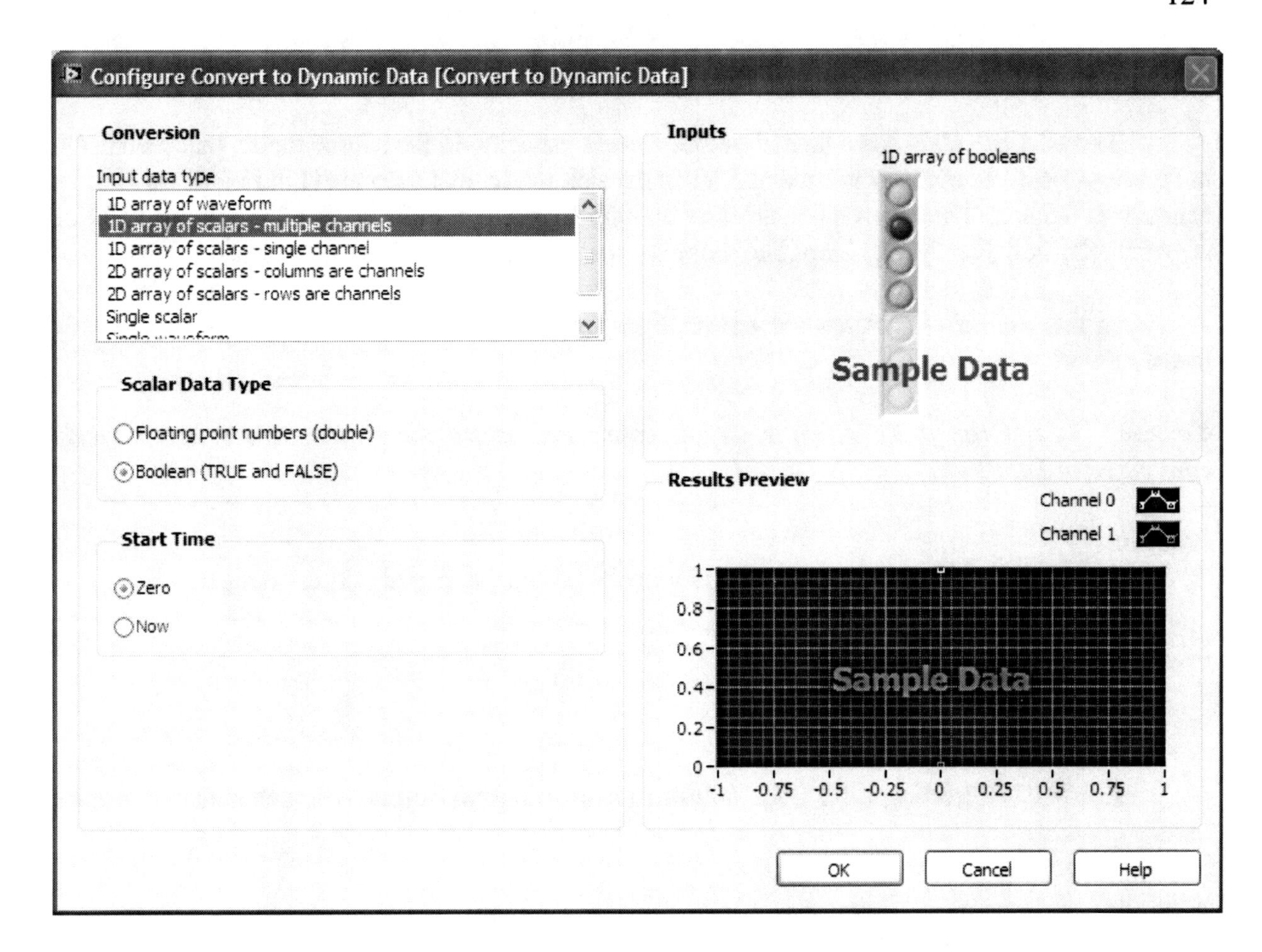

Figure 4-4-6-1-5 "Configure Convert to Dynamic Data" window.

14) Run the VI. Pressing the pushbuttons should cause the "Round LED"s in the "Front Panel" to go out.

4.4.6.2 Creation of Digital Output Voltages

The NI USB-6009 has a limited output current capacity. In the source mode, that seen in (2) in Figure 4-4-6-1, its capacity is 1 ma at 5 V. In the sink mode, that seen in (1) in Figure 4-4-6-1, its capacity is 8.5 ma. These capacities are low enough to make it difficult to fully light an LED or power a relay coil directly. Op Amps like those shown in Figure 4-4-6-1 are usually needed.

In this example Op Amps are not necessary since high-impedance voltmeters are the only loads.

Caution: Do not run a VI with the circuit connected before the digital inputs are properly configured in LabVIEW.

Problem:

Use LabVIEW and the NI USB-6009 to make switchable digital output voltages.

Equipment:

NI USB-6009

Two VOMs set to read DC voltage

LabVIEW directing a NI USB-6009 to produce digital output voltages is demonstrated here.

Solution:

1) Build the circuit of Figure 4-4-6-2-1.

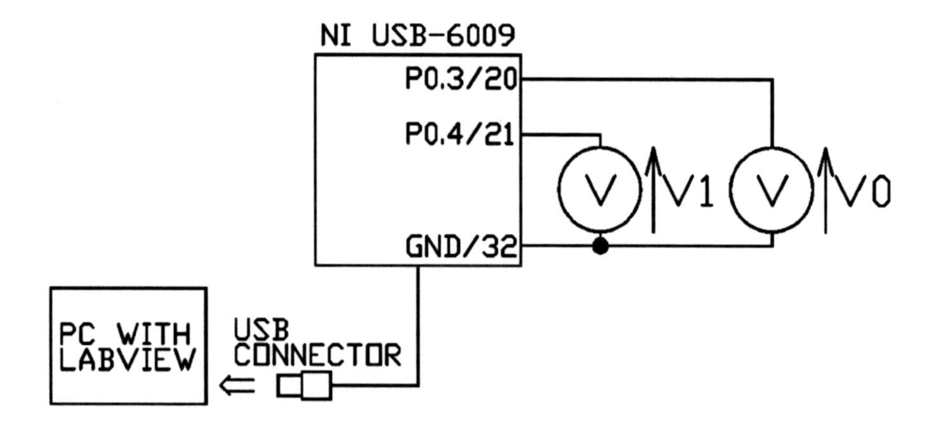

Figure 4-4-6-2-1 Two voltmeters reading digital output data from the NI USB-6009.

2) Create the VI of the “Front Panel” of Figure 4-4-6-2-2 and the “Block Diagram” of Figure 4-4-6-2-3.

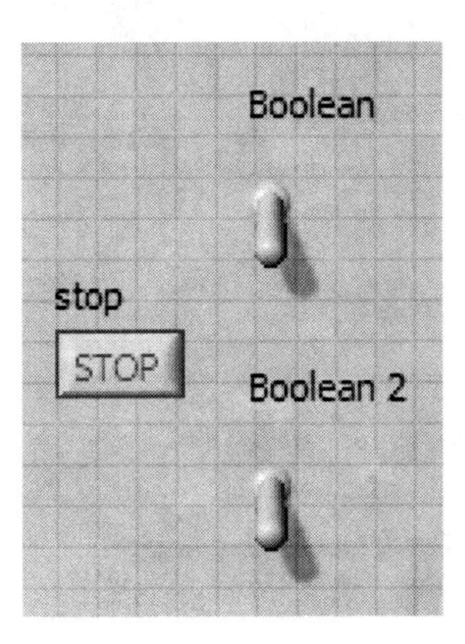

Figure 4-4-6-2-2 “Front Panel” with “Vertical Toggle Switch” input blocks for .controlling the NI USB-6009 output.

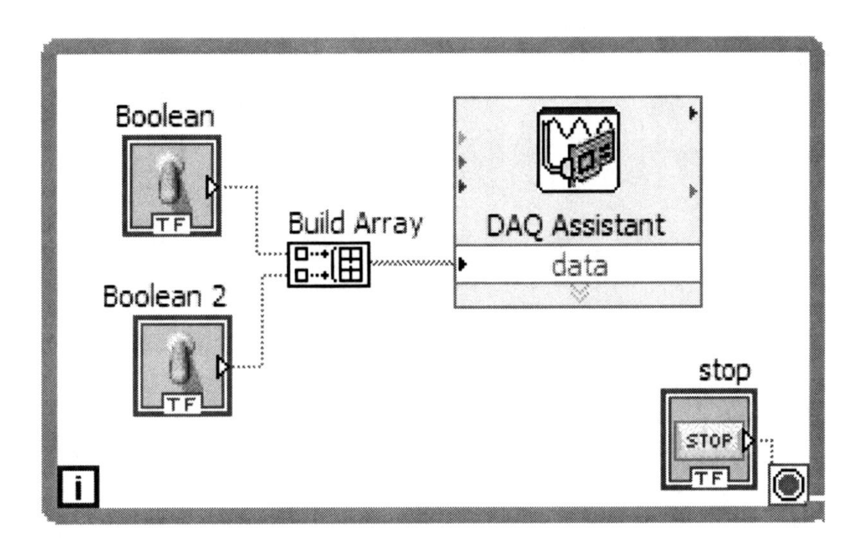

Figure 4-4-6-2-3 “Block Diagram” for producing digital output voltages.

3) To create the “DAQ Assistant“ function block left-click on “View”, “Functions Palette”, and “Output”. Then left-click, hold, and drag it onto the “Block Diagram”.

4) In the appearing window left-click on “Generate Signals”, “Digital Output”, “Line Output”, “Dev2 (USB-6009)”, “port0/line3”, “Ctrl” “port0/line4” and “Finish”.

5) In the appearing "DAQ Assistant" window leave the default values. See Figure 4-4-6-2-4.

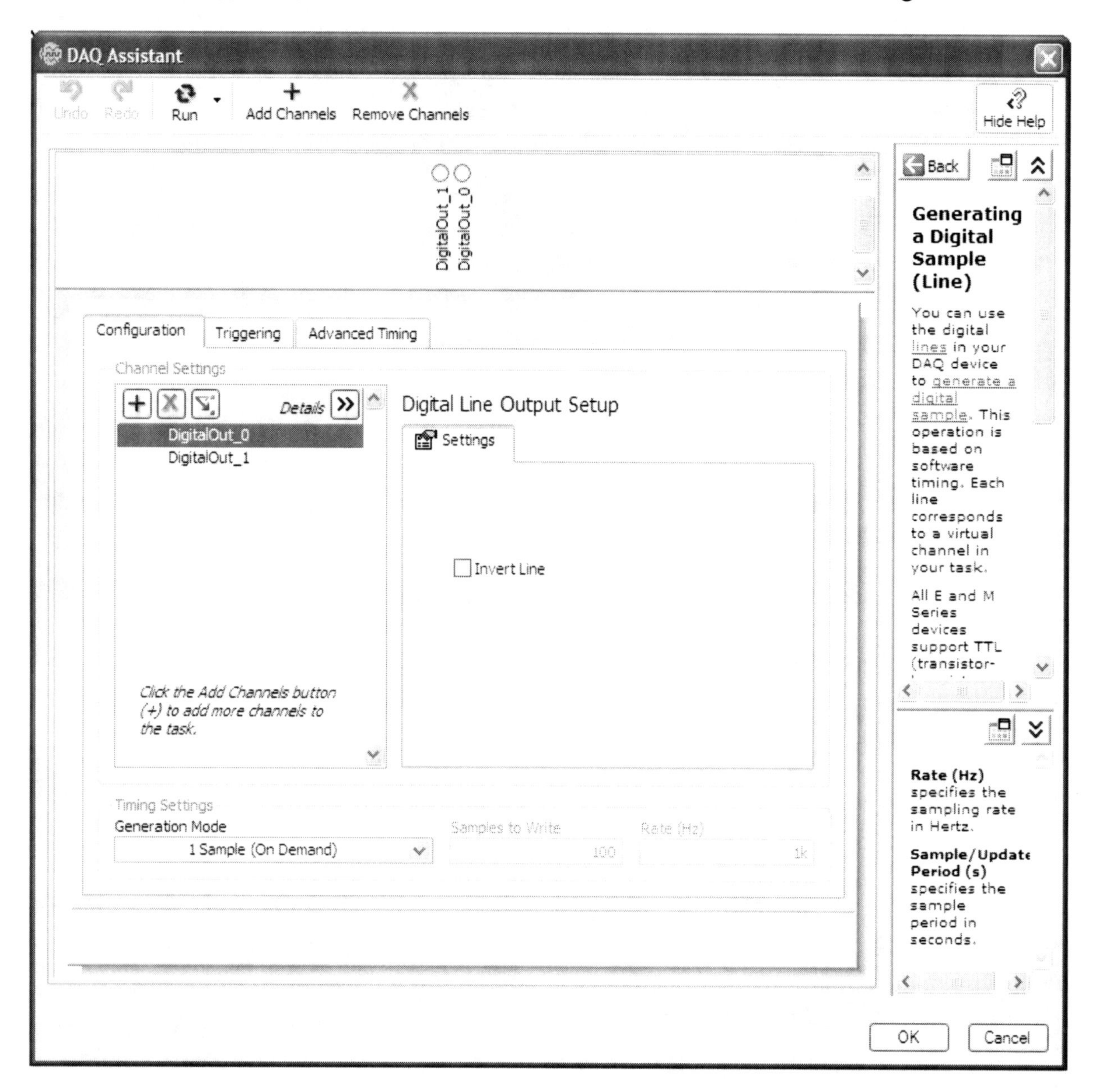

Figure 4-4-6-2-4 "DAQ Assistant" window for digital output.

6) Still in the "DAQ Assistant" window, left-click on "Run". By left-clicking the little circles above "DigitalOut_0" and "DigitalOut_1" a green dot appears or disappears in the little circles and the digital outputs at P03 and P04 produce 5 V or 0 V across the voltmeter. This "Run" is for testing the NI USB-6009 and digital output circuit. When the VI is run, the voltage is controlled from the "Front Panel" as shown in Figure 4-4-6-2-2.

7) Left-click "OK" at the bottom of the "DAQ Assistant" window.

8) Put a "While Loop" on the "Front Panel" by left-clicking on "View", "Functions Palette", "Express", and "Execution Control". Drag the "While Loop" onto the "Front Panel" and stretch it out.

9) Drag the "DAQ Assistant" function block into the "While Loop".

10) On the "Front Panel" left-click on "View", "Controls Palette", "Express", "Buttons & Switches". Then left-click and drag a "Vertical Toggle Switch" input block onto the "Front Panel". Repeat this to drag on a second "Vertical Toggle Switch" input block.

11) The "Vertical Toggle Switch" input blocks produce Boolean values, but the "DAQ Assistant" function block, as configured here, requires a one dimensional Boolean array. A "Build Array" function block is used to make the conversion. Get it by left-clicking on "View", "Functions Palette", "Programming", and "Array". Drag the "Build Array" function block into the "While Loop". Add another input to the "Build Array" function block by right-clicking on it and left-clicking on "Add Input".

12) Connect the blocks by dataflow wires as in Figure 4-4-6-2-3.

13) Run the VI. Left-click toggling the "Vertical Toggle Switch" input blocks should cause the VOM voltmeters to see either 0 or 5 V.

4.4.7 COUNTER

Using the PFI0 line, the NI USB-6009 can count falling voltage edges. It then stores the counts in a 32-bit counter.

DAQs' counters are generally fast devices that are sensitive to make/break voltage spikes like those that are caused by bouncing mechanical switches. Some NI DAQs that have a built in digital filtering feature, but the NI USB-6009 does not.

When a mechanical switch or pushbutton opens or closes it may make and break electrical contact several times before reaching a definite steady opened or closed state. This would make the PFI0 counter receive false counts. To correct this, a debouncing circuit is used to keep the PFI0 from receiving the false make/break counts.

There are simple passive RC debouncing circuits and better performing, but more complicated, active electronic debouncing circuits. A simple RC debouncing circuit is demonstrated here. More details on debouncing can be seen in literature and on the internet. The National Instruments "KnowledgeBase" web pages are a good source.

Caution: As with the digital inputs, do not run a VI with the circuit connected before the counter input is properly configured in LabVIEW.

Problem:

Use LabVIEW and the NI USB-6009 to count the number of times a pushbutton is closed. Do this with and without a simple RC debouncing circuit.

Equipment:

NI USB-6009
Pushbutton
4.7 kΩ resistor
10 Ω resistor
47 μF capacitor

NI USB-6009's inputting its counter output into LabVIEW and a simple RC debouncing circuit are demonstrated here.

Solution:

1) Build the debounced circuit of Figure 4-4-7-1.

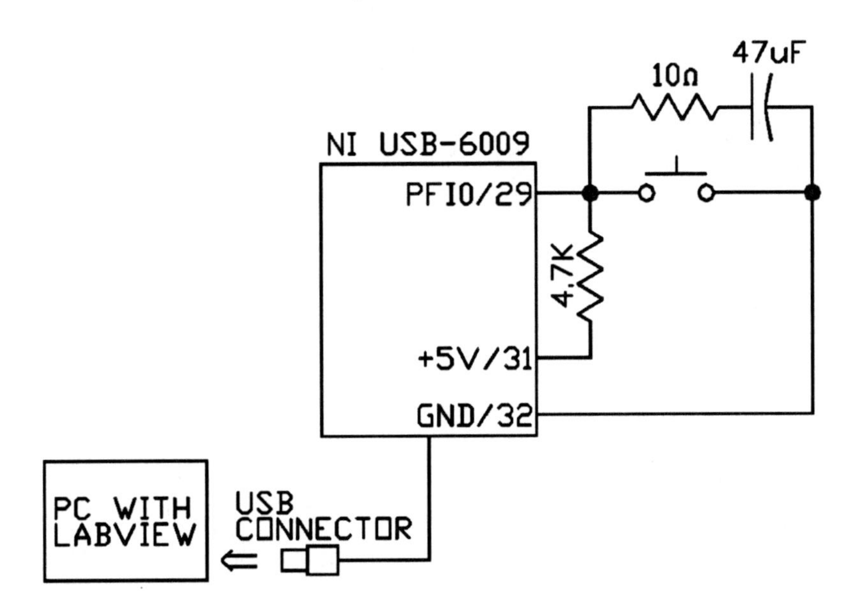

Figure 4-4-7-1 NI USB-6009 used as pushbutton position counter.

2) Create the VI of the "Front Panel" of Figure 4-4-7-2 and the "Block Diagram" of Figure 4-4-7-3.

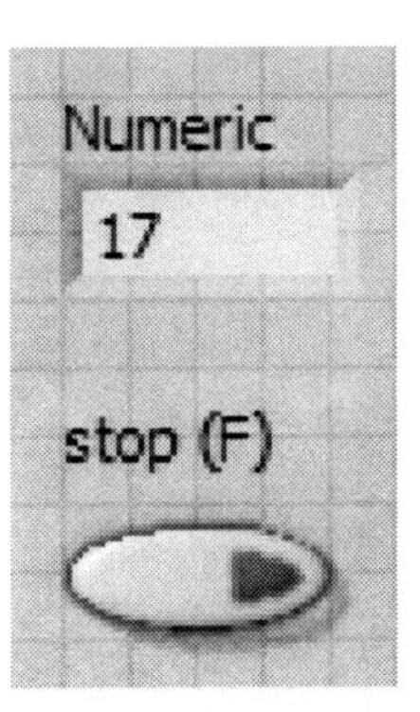

Figure 4-4-7-2 "Front Panel" showing the number of times the PFI0 counter has seen voltage drop from +5 V to 0 V.

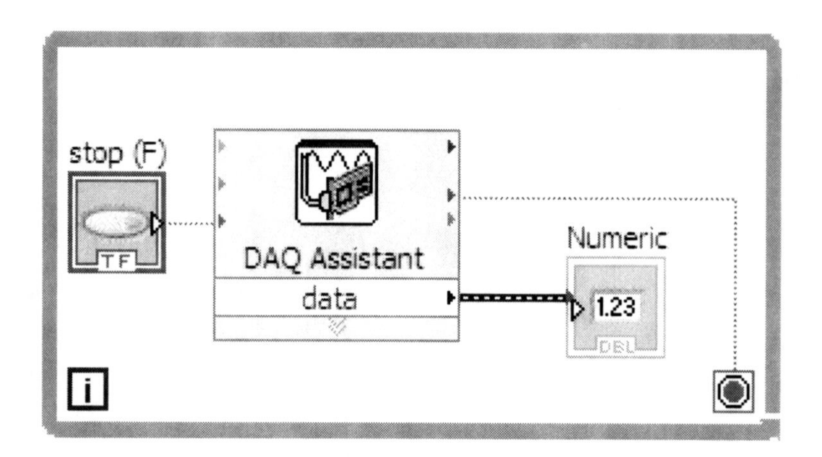

Figure 4-4-7-3 "Block Diagram" for counting the number of times the PFI0 counter has seen voltage drop from +5 V to 0 V.

3) To create the "DAQ Assistant" function block left-click on "View", "Functions Palette", "Express", and "Input". Then left-click, hold, and drag it onto the "Block Diagram".

4) In the appearing window left-click on "Acquire Signals", "Counter Input", "Edge Count", "Dev2 (SUB-6009)", "ctr0", and "Finish".

5) In the appearing "DAQ Assistant" window leave the default values. See Figure 4-4-7-4.

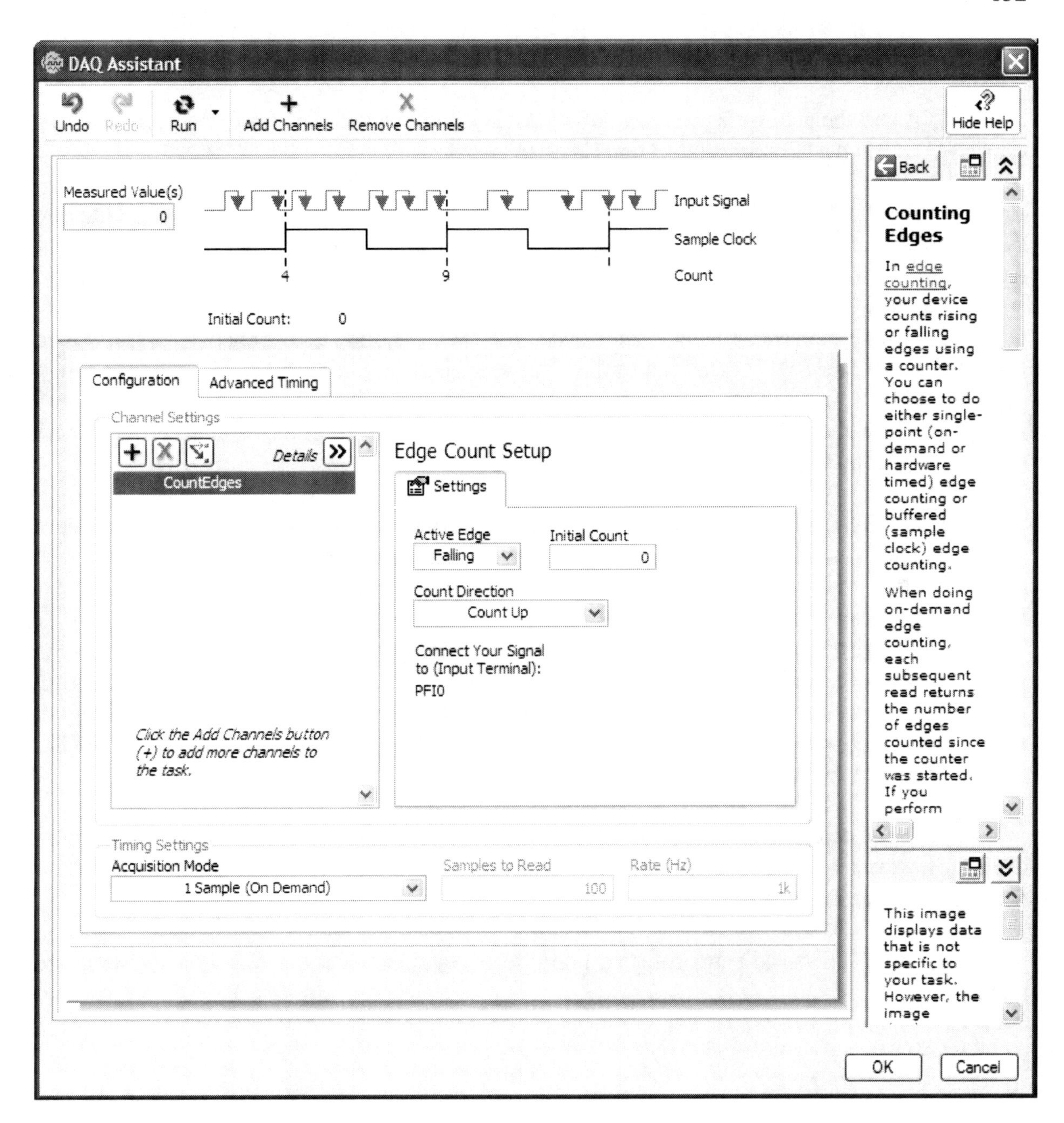

Figure 4-4-7-4 "DAQ Assistant" window for the counter input.

6) Still in the "DAQ Assistant" window, left-click on "Run". This starts the counter. Each time the pushbutton is pushed the "Measured Value(s)" count (in the small window in the upper left of the "DAQ Assistant" window) should increase by 1. See the Results Discussion at the end of this example. This "Run" is for testing the NI USB-6009 and switch circuit. When the VI is run, the count will appear in the "Front Panel" of Figure 4-4-7-2.

7) Left-click "OK" at the bottom of the "DAQ Assistant" window.

8) Allow the automatic placement of a "While Loop" around the "DAQ Assistant" by left-clicking "Yes" on the "Confirm Auto Loop Creation" window.

9) On the "Front Panel" left-click on "View", "Controls Palette", "Express", "Numeric Indicators". Then left-click and drag a "Numeric Indicator" output block onto the "Front Panel".

10) Move the "Numeric" indicator inside the "While Loop" and connect a dataflow wire from the "DAQ Assistant" "data" out to the "Numeric" indicator's input.

11) Run the VI. Each time the pushbutton is pressed the count in the "Front Panel" should increase by 1. See the following Results Discussion.

12) Stop the VI, remove the RC circuit, and rerun the VI. Note, as expected, there are more extra counts without the RC circuit. See the following Results Discussion.

Results Discussion:

The pushbutton used when writing this book did not always make good contact. When the switch made poor, make/break, contact, extra voltage edges were created. Then the "DAQ Assistant" function block counted extra counts. The RC debounce circuit improved the counting, but did not completely solve it. False counts still occurred after the approximate .1 seconds that the RC circuit debounced the switch.

A larger capacitance in the RC circuit would have further decreased the unwanted extra counts. However, a larger capacitance would increase the system's needed response time and so decrease its maximum counting rate.

To avoid unwanted extra counts a good debouncing circuit and a high quality switch are required.

5.0 REFERENCES

1) The LabVIEW 2009 Student Edition Textbook Bundle includes the LabVIEW Student Edition software and Dr. Robert H. Bishop's popular introductory textbook *Learning with LabVIEW*, published by Prentice Hall. The text book is 752 pages long and 9" x 7.3". With the software, the bundle's list price is $115.00.

2) Essick, John, *Hands-On Introduction to LabVIEW for Scientists and Engineers*, Oxford University Press, 2009. It is paperback, 511 pages long, and 9.2" x 7.5". Its list price is $31.95. Essick wrote this using his years of experience teaching LabVIEW at Reed College. It emphasizes LabVIEW programming but also contains information on real-world projects. The book is wordy, but good.

3) Travis, Jeffery and Kring, Jim, *LabVIEW for Everyone: Graphical Programming Made Easy and Fun*, 3rd Edition, Pearson Education, Inc, 2007. It is hardcover, 981 pages long, and 9.2" x 7". It includes a CD with example code and an evaluation copy of LabVIEW version 8.0. Its list price is $81.99. Travis and Kring wrote this using their years of teaching and industrial LabVIEW experience. It emphasizes LabVIEW programming but also provides some details on connecting a computer running LabVIEW through data acquisition and control devices to real equipment.

4) Blume, Peter A., *The LabVIEW Style Book*, Prentice Hall, 2007. It is hardcover, 372 pages long, and 8.25" x 10.25". Its list price is $104.00. This is a popular book. Its author says that his book is meant for experienced and advanced beginner LabVIEW programmers. The book would be of greatest use to those wishing to write large checkable VIs that are understandable to other people.

5) *LabVIEW, Getting Started with LabVIEW*, National Instruments, 2009, Part Number 37327F-01. It is 90 pages long and 8 ½" x 11". Copies are available for free from National Instruments on their website http://digital.ni.com/manuals.nsf/websearch/D27DC92E8D6556CD862575AC0074EDAB. This is a tutorial on LabVIEW graphical programming, data acquisition, and instrument control.

6) *LabVIEW User Manual*, National Instruments, 2003, Part Number 320999E-01. It is 349 pages long and 8 ½" x 11". Copies are available for free from National Instruments on their website http://digital.ni.com/manuals.nsf/websearch/790127B60590AD0C86256D2C005DCE0F?OpenDocument&seen=1 This manual describes the LabVIEW graphical programming environment and techniques for building applications in LabVIEW, such as test and measurement, data acquisition, instrument control, datalogging, measurement analysis, and report generation applications.

7) *LabVIEW Measurements Manual*, National Instruments, 2003, Part Number 322661B-01. It is 159 pages long and 8 ½" x 11". Copies are available for free from National Instruments on their website http://digital.ni.com/manuals.nsf/websearch/D27DC92E8D6556CD862575AC0074EDAB It is a supplement to the *LabVIEW Users Manual*. It contains information that is useful in acquiring and analyzing measurements data.

8) *Introduction to LabVIEW and Computer-Based Measurements Hands-On Seminar*, National Instruments, Part Number 342377F-01, January 2010. It is a spiral bound manual, 100 pages long, and 8.5" x 11". Presently, copies of this are distributed during free 3 hour NI seminars. At the time of writing, information on these seminars can be found on the website http://sine.ni.com/apps/utf8/nievn.ni?action=display_offerings_by_event&event_id=26298®ion=ne&site=NIC&node=61110&l=US If possible, the beginner should attend one of these seminars and obtain the manual.

9) *User Guide and Specifications NI USB-6008/6009*, National Instruments, Part Number 371303L-01, It is 32 pages long and 8 ½" x 11". Copies are available for free from National Instruments on their website http://www.ni.com/pdf/manuals/371303l.pdf.

6.0 APPENDIX

6.1 TOOLS PALETTE CURSORS

The "Tools Palette" window can be brought up by left-clicking on "View" and "Tools Palette". Then left-click on the icons in it to create different cursors.

The cursors used in this book are:

"Connect Wire" This looks like a wire spool. It is used for connecting and drawing dataflow wires.

"Edit Text" This is a large A. It is used for writing and editing text.

"Operate Value" This looks like a hand. It is used for selecting and modifying numerical values.

"Position/Size/Select" This is an arrow. It is used for positioning blocks, sizing objects, and selecting blocks and objects.

6.2 LabVIEW DATAFLOW WIRES

LabVIEW uses different shaped and colored dataflow wires for different types of data. The back cover shows common default dataflow wires that were used in this book. LabVIEW is also capable of creating custom shaped and colored dataflow wires.

6.3 NUMERIC PRECISION

By default, LabVIEW stores all numeric values as floating point and double-precision (approximately 16 decimal digits). It is possible to specify lesser precision.

By default, LabVIEW displays 5 or 6 digits. The number of displayed digits can be adjusted.

6.4 TIPS

Save VIs soon and often.

To more easily see what function blocks are available, from the “Block Diagram” left-click on “View”, “Functions Palette”, “View” <in the palette heading>, “View This Palette As”, and “Category (Icons and Text)”.

Change from “Front Panel” to “Block Diagram” or vice versa by typing “Ctrl E”.

To display the “Front Panel” and “Block Diagram” at the same time left-click on “Window” and “Tile Left and Right” or type “Ctrl T”.

An effort should be made to make VIs as clear and understandable as possible. To do this align and locate blocks logically, use “Formula Node”s for equations, and use the MathScript add-on for VIs that contain many equations.

If you are having trouble writing a VI, get a small part of it running and then build on that.

Examples of VIs can be found in the “NI Example Finder”. To get to it, left-click on “Help” and “Find Examples”.

To get help on a block, right-click on it and left-click on “Help”.